Energy Science, Engineering and Technology

Battery Technology Advancements in China

ENERGY SCIENCE, ENGINEERING AND TECHNOLOGY

Additional books in this series can be found on Nova's website
under the Series tab.

Additional E-books in this series can be found on Nova's website
under the E-books tab.

MANUFACTURING TECHNOLOGY RESEARCH

Additional books in this series can be found on Nova's website
under the Series tab.

Additional E-books in this series can be found on Nova's website
under the E-books tab.

ENERGY SCIENCE, ENGINEERING AND TECHNOLOGY

BATTERY TECHNOLOGY ADVANCEMENTS IN CHINA

JOHN S. GAGLIONE
EDITOR

Nova Science Publishers, Inc.
New York

Copyright © 2012 by Nova Science Publishers, Inc.

All rights reserved. No part of this book may be reproduced, stored in a retrieval system or transmitted in any form or by any means: electronic, electrostatic, magnetic, tape, mechanical photocopying, recording or otherwise without the written permission of the Publisher.

For permission to use material from this book please contact us:
Telephone 631-231-7269; Fax 631-231-8175
Web Site: http://www.novapublishers.com

NOTICE TO THE READER

The Publisher has taken reasonable care in the preparation of this book, but makes no expressed or implied warranty of any kind and assumes no responsibility for any errors or omissions. No liability is assumed for incidental or consequential damages in connection with or arising out of information contained in this book. The Publisher shall not be liable for any special, consequential, or exemplary damages resulting, in whole or in part, from the readers' use of, or reliance upon, this material. Any parts of this book based on government reports are so indicated and copyright is claimed for those parts to the extent applicable to compilations of such works.

Independent verification should be sought for any data, advice or recommendations contained in this book. In addition, no responsibility is assumed by the publisher for any injury and/or damage to persons or property arising from any methods, products, instructions, ideas or otherwise contained in this publication.

This publication is designed to provide accurate and authoritative information with regard to the subject matter covered herein. It is sold with the clear understanding that the Publisher is not engaged in rendering legal or any other professional services. If legal or any other expert assistance is required, the services of a competent person should be sought. FROM A DECLARATION OF PARTICIPANTS JOINTLY ADOPTED BY A COMMITTEE OF THE AMERICAN BAR ASSOCIATION AND A COMMITTEE OF PUBLISHERS.

Additional color graphics may be available in the e-book version of this book.

Library of Congress Cataloging-in-Publication Data

Battery technology advancements in China / editor, John S. Gaglione.
    p. cm.
Includes index.
ISBN 978-1-61470-563-5 (hardcover)
1. Electric batteries--China. I. Gaglione, John S.
TK2901.B385 2011
621.31'242--dc23
                            2011024372

Published by Nova Science Publishers, Inc. † New York

CONTENTS

Preface   vii

Chapter 1   Developments in Lithium-Ion Battery Technology in the People's Republic of China   **1**
Argonne National Laboratory and Energy Systems Division

Chapter 2   Advanced Battery Technology for Electric Two-Wheelers in the People's Republic of China   **75**
Argonne National Laboratory and Energy Systems Division

Index   **231**

PREFACE

Lithium-ion battery technology is widely used in portable electronics products, such as cell phones, camcorders and portable computers. At present, this technology is gaining worldwide attention as a battery option for transportation applications, including electric vehicles, hybrid vehicles, plug-in hybrid vehicles, fuel cell vehicles and electric bikes in Asia. The United States continues to lead in the research and development of lithium-ion technology. However, in Asia, countries like China are commercializing and producing this technology. This book examines battery technology advancements in China where there are over 120 companies involved in the production of lithium-ion battery technology.

Chapter 1- Lithium-ion battery technology is widely used in portable electronics products, such as cell phones, camcorders, and portable computers. At present, this technology is gaining worldwide attention as a battery option for transportation applications, including electric vehicles (EVs), hybrid vehicles, plug-in hybrid vehicles (PHEVs), fuel cell vehicles, and electric bikes in Asia.

Chapter 2- This report focuses on lithium-ion (Li-ion) battery technology applications for two- and possibly three-wheeled vehicles. The author of this report visited the People's Republic of China (PRC or China) to assess the status of Li-ion battery technology there and to analyze Chinese policies, regulations, and incentives for using this technology and for using two- and three- wheeled vehicles. Another objective was to determine if the Li-ion batteries produced in China were available for benchmarking in the United States.

Chapter 1

DEVELOPMENTS IN LITHIUM-ION BATTERY TECHNOLOGY IN THE PEOPLE'S REPUBLIC OF CHINA

Argonne National Laboratory and Energy Systems Division

NOTATION

The following is a list of the abbreviations, acronyms, and units of measure used in this document. (Some acronyms and abbreviations used only in tables may be defined only in those tables.)

GENERAL ACRONYMS AND ABBREVIATIONS

ABAT	Advanced Battery Technologies, Inc.
ASME	American Society of Mechanical Engineers
BIT	Beijing Institute of Technology
CATARC	China Automotive Technology and Research Center
CE	Conformité Européenne (French for "European Conformity")
DOD	depth of discharge
DOE	U.S. Department of Energy
EV	electric vehicle
GB	Guobiao
GM	General Motors (Corporation)
GRINM	General Research Institute for Nonferrous Metals
HEV	hybrid electric vehicle
IEC	International Electrical Commission
MGL	Mengguli Corporation (MGL)
MOST	Ministry of Science and Technology
NLI	nano-lithium-ion

NREL	National Renewable Energy Laboratory
OVT	Office of Vehicle Technologies Program
PHEV	plug-in hybrid electric vehicle
PLI	polymer lithium-ion
PRC	Peoples Republic of China
PV	photovoltaic
QSTC	Chemical and Physical Power Sources of China Ministry of Information Industry
R&D	research and development
RMB	Ren Min Bi
SEI	solid-electrolyte interface
SOC	state of charge
UL	Underwriters Laboratories
USABC	U.S. Advanced Battery Consortia
VRLA	valve-regulated lead-acid

Other Abbreviations

Li-ion	lithium-ion
$LiCoO_2$	lithiated cobalt-oxide
$LiFePO_4$	LFP; lithiated iron phosphate (olivine)
$LiMnO_2$	LMS; lithium manganese spinel
$Li(Ni_{0.85}Co_{0.1}Al_{0.05})O_2$	NCH; lithiated mixed oxide of nickel, cobalt, and aluminium
$Li(Ni_{1/3}Co_{1/3}Mn_{1/3})O_2$	NCM; lithiated mixed oxide of nickel, cobalt, and manganese
NiCd	nickel cadmium
NiMH	nickel metal hydride

ACKNOWLEDGMENTS

My heartfelt thanks go to Jinhua Zhang, Vice President; Hou Fushen, Director, Hi-tech Development Department; and Zhao Chunming, Master Senior Engineer, all from the China Automotive Technology & Research Center (CATARC), who provided first-rate help and assistance with all the logistics involved in arranging meetings and moving to and from meetings in a timely manner, making my visit to the Peoples Republic of China a success. Many thanks.

I would like to thank our sponsor, Tien Duong, Team Leader, Hybrid and Electric Systems, Vehicles Technologies Program, Office of Energy Efficiency and Renewable Energy, for his support and encouragement for undertaking this visit to study lithium-ion battery technology in China.

Also, I very much appreciate the efforts of Dr. Larry R. Johnson, Director of the Transportation Technology Research and Development Center at Argonne National Laboratory, and my colleagues, Dr. Michael Q. Wang and Dr. Danilo J. Santini, toward guiding me to make this visit a success and preparing me to conduct this lithium-ion battery study.

I also thank Kevin A. Brown, Argonne's Technical Services Division (TSD), for his help in preparing and coordinating the production of this report, and Vicki Skonicki, also of TSD, for document production.

SUMMARY

Lithium-ion battery technology is widely used in portable electronics products, such as cell phones, camcorders, and portable computers. At present, this technology is gaining worldwide attention as a battery option for transportation applications, including electric vehicles (EVs), hybrid vehicles, plug-in hybrid vehicles (PHEVs), fuel cell vehicles, and electric bikes in Asia.

The United States continues to lead in the research and development of lithium-ion technology. There is a strong research and development program funded by the U.S. Department of Energy and other federal agencies, such as the National Institute of Standards and Technology and the U.S. Department of Defense. However, in Asia, countries like China, Korea, and Japan are commercializing and producing this technology. China has over 120 companies involved in the production of lithium-ion battery technology. The following are highlights and conclusions based on the information provided in this report:

- Lithium-ion batteries offer higher power and energy per unit weight and volume and better charge efficiency than nickel metal hydride (NiMH) batteries. These attributes allowed them to capture a major part of the portable rechargeable battery market within a few years of their introduction and to generate global sales estimated at $5 billion in 2006.
- Battery temperature affects battery performance and life. Therefore, battery thermal management is critical to achieving desired performance and calendar life for battery packs in hybrid electric vehicles. Automakers and their suppliers are paying increased attention to battery thermal management to ensure that warranty costs for battery replacement do not exceed projections. The battery in a hybrid electric vehicle (HEV) experiences a demanding thermal environment and must be heated during cold-weather operation and cooled during extended use and during warm-weather operation.
- Mass-manufacturers of lithium-ion cells for consumer products are now engaged in the development of lithium-ion chemistries for HEV and PHEV applications, with commercialization possible as early as 2010. The major impediment to developing lithium-ion batteries for PHEVs appears to be that the PHEV battery requirements are not clearly defined at this time.
- Battery thermal management is critical to achieving performance and extended battery life in electric and hybrid vehicles under real driving conditions. To find a high-performance and cost-effective cooling system, system thermal response and its sensitivity must be evaluated as a function of controllable system parameters.

- A well-designed thermal management system is required to regulate electric vehicle (EV) and HEV battery pack temperatures evenly, keeping those within the desired operating range. Production HEVs require an active heating/cooling battery temperature monitoring system to allow them to operate in hot and cold climates. Proper thermal design of a module has a positive impact on overall pack thermal management and its behavior.
- A key attraction of lithium-based batteries is the high cell voltage, which is the direct result of the highly negative potential of lithium. With currently used mixed oxide positives, the operating voltage range of lithium-ion cells is approximately 2.75 to 4.2 V. The nominal (average) discharge voltage is about 3.6 V, and most of the usable cell capacity is delivered between 4.0 and 3.5 V.
- The calendar life of high-power lithium-ion battery cells is expected to have the same basic dependence on temperature as high-energy cell designs, because several of the high-power cell technologies use the same basic chemistry as larger cells and thus are subject to the same kind of degradation processes.
- The current cost of lithium-ion HEV batteries made with early pilot equipment is around $2,000/kWh. If the current designs were brought to volume production, the cost is anticipated to drop to approximately $1,000/kWh.
- In the past three to four years, companies outside of the Peoples Republic of China (PRC) have been bringing advanced battery technologies to the PRC and setting up partnerships and/or joint ventures to manufacture batteries for these and other applications (such as electric bikes, EVs, and HEVs) to take advantage of low labor cost and incentives provided by the Chinese government. Companies in the PRC are very aggressive in developing manufacturing processes for the batteries export market.
- From 2001 to 2004, the number of battery companies in China increased from 455 to 613; accordingly, the number of employees in those industries also increased from 140,000 in 2001 to 250,000 in 2004. The total output reached 63.416 billion Yuan ($8.1 billion) in 2004, which is an increase of 52.58% over 2001.
- The sales of large-scale companies in the battery industry was 59.818 billion Yuan ($7.65 billion) in 2004 — this was an increase of 52.85% in comparison with 2003, an increase of 105.32% in comparison with 2002, and an increase of 160.93% in comparison with 2001. This growth is attributed to the growth of large companies. In the last four years, the debt-to-asset ratio of China's battery industry has been fluctuating between 54 and 59%.
- The most commonly used battery industry standards in China for testing and evaluating battery technologies are those from the International Electrical Commission (IEC).
- Along with the rapid growth of lithium-ion battery manufacturers in China, companies like the BYD Company Limited; Tianjin Lishen Battery Joint- Stock Co., Ltd.; Shenzhen BAK Battery Co., Ltd.; and Shenzhen B&K Technology Co., Ltd., are increasing their share of the market. In 2004, the domestic and overseas markets for lithium-ion batteries were flourishing — the export volume was 189 million units, with an increase of 16.3% in sales. As a result of the rapid increase in domestic

- demand, the import volume of lithium-ion batteries was 550 million units, with an increase of 23.43% in sales in 2004.
- At present, Chinese lithium-ion battery manufacturing companies are relatively well developed. Such manufacturers as the BYD Company Limited; Shenzhen BAK Battery Co., Ltd.; and Shenzhen B&K Technology Co., Ltd., enjoy a large share of the global battery market. During 2003–2004, the Chinese lithium-ion battery industry developed dramatically. The production of cobalt acid lithium and nickel acid lithium and the invention of new manufacturing techniques to extract lithium from salty lakes will drastically reduce the need to import anode materials for lithium batteries from abroad.
- Most Chinese companies are producing lithium-ion batteries for portable applications. Large companies have undertaken research and development with the help of joint ventures and/or partnerships with companies from Japan, Europe, and the United States.
- Lithium resources are abundant in China. As of 2000, China was the second largest producer of lithium in the world, and in 2004, it produced 18,000 metric tons.
- Lithium-ion battery packs for e-bikes range from 24 to 37 V and have a capacity of 5–60 A•h. The market for lithium-ion e-bikes in China is still small. In Japan and Europe, however, lithium-ion and NiMH are the dominant battery types, although annual e-bike sales (200,000/yr and 100,000/yr, respectively) are significantly less than those in China.
- Tsinghua University's focus is research on EVs, HEVs, PHEVs, and fuel cell electric vehicles (FCEVs), with an emphasis on battery applications. The university has six patents and four applications pending in China on battery thermal management, EV controller design, and electronics for vehicles. University staff is interested in developing a relationship with a U.S. bus manufacturer, fuel cell developer, and hydrogen storage company.
- CITIC GUOAN Mengguli Corporation (MGL); MGL New Energy Technology Co., Ltd.; Green Power Co.; Tianjin Lintian Double-Cycle Tech. Co.; Suzhou Phylion Battery Co., Ltd.; Tianjin Lishen Battery Joint-Stock Co., Ltd.; and Tongji University are willing and interested in providing cells, modules, and packs for benchmarking in the United States. However, these organizations would want to see a test plan up front and would like to keep their test data out of the public domain.
- The China Electrotechnical Society has 123,000 members. The society is a clearinghouse for electrotechnical research. It conducts studies on battery technology markets for EVs and HEVs. Recently, it completed a preliminary technology study on fuel cells and fuel cell hybrid vehicles for its membership and the Chinese government. It was learned that the Chinese government emphasizes the development of lithium-ion battery technology for vehicular applications. The Chinese government is providing incentives and grants for Chinese-owned and Chinese-operated companies — amounting to as much as 75% of the total cost, which covers most operating and capital expenses. For small companies, these incentives can be upwards of 85%.
- MGL ranks as China's largest manufacturer of the lithium-ion battery cathode material $LiCoO_2$, and it will be the first to market the new cathode materials

$LiMn_2O_4$ and $LiCoO_{0.2}Ni_{0.8}O_2$. MGL emphasizes quality control, and has passed the certification of both New and Hi-Tech Enterprise standards and ISO9001:2000. With its own synthesis method, MGL claims it produces cathode materials of superior performance and reliability in an environmentally friendly way. Since incorporation, MGL has held a monopoly in China's lithium-ion battery cathode materials market and now is at the forefront of the industry. Besides cathode materials, MGL also produces lithium-ion secondary batteries of high energy density and high capacity for power and energy storage — the capacity ranges from several ampere-hours to several hundred ampere-hours. As China's first power battery manufacturer, MGL leads in marketing high-capacity lithium-ion secondary batteries, which are used in the Beijing Municipality's trial electric bus fleet.

- Tianjin Institute of Power Sources is one of the two national laboratories involved in battery testing and evaluation activities and programs. This testing center of chemical and physical power sources of the Ministry of Information Industry of China was established in 1985. It is considered the largest, most comprehensive, most authoritative, independent quality-testing center for chemical and physical power sources.
- Tianjin Lishen Battery Joint-Stock Co., Ltd., was established in 1998. It has a capitalization of 600 million Ren Min Bi (RMB) ($80.00 million), and a total investment of 1.5 billion RMB ($200.00 million). The production of lithium cells is completely automatic — representing the most automated production line for lithium-ion batteries in China. The production equipment is imported from Japan.
- At its School of Automotive Engineering, Tongji University has world-class facilities to integrate advanced batteries and fuel cells in vehicles and to conduct basic and applied research for the automotive industry. These testing capabilities cover research, testing, and evaluation. The school is collaborating research with lithium battery development companies, fuel cell development companies, and domestic and foreign automobile companies.
- Suzhou Phylion Battery Co., Ltd., is a battery technology corporation set up by Legend Capital Co., Ltd.; the Institute of Physics of the Chinese Academy of Sciences; and Chengdu Diao Group. Suzhou Phylion Battery Co., Ltd., has 82 million RMB ($10.93 million) and a staff of more than 400. The company specializes in manufacturing and selling lithium-ion cells with high capacity and current. Its technology is primarily used in defense, electric bicycles, lighting, portable electronics, medical equipment, and battery-operated tools.
- The General Research Institute for Nonferrous Metals (GRINM), established in 1952, is the largest research and development (R&D) institution in the field of nonferrous metals industry in China. GRINM is conducting basic research on the materials needs and requirements for high-energy and high-power lithium-ion battery technology. GRINM has focused on nanotechnology and $LiMn_2O_4$ materials for the cathode, graphite for the anode, and PC+DC+DMC+1m Li P F6 liquid electrolyte and polypropylene/ polyethylene/polypropylene separator for the development of a lithium-ion battery cell. GRINM developed all materials in-house except for the separator, which GRINM imports from Japan and the United States.
- The production and manufacturing of lithium-ion batteries has expanded enormously in the past two to three years. These batteries are widely used in consumer

electronics devices, such as cellular phones, camcorders, and portable computers. This market is growing fast and is second only to that of information technology. At companies we visited, we saw batteries manufactured by a wide variety of processes — from manual production to full automation. Labor cost also plays a significant role in manufacturing lithium-ion batteries. Currently, China has between 275,000 and 325,000 workers in the battery industry. Labor costs in China are very low compared with those in Japan and western countries. Some 435 batteries companies in China are producing batteries. In 2006, their total output was 78.7 billion Yuan.

- Western and Japanese companies are setting up joint ventures and/or partnerships with Chinese battery companies, taking advantage of widespread incentives available to Chinese companies. Chinese companies receive more incentives in producing batteries for export. Today, Chinese companies are providing batteries for electronic and portable applications worldwide. Because the initial investment to produce lithium-ion batteries is very high, producing them in China makes sense for Japanese and western companies.
- The rechargeable lithium battery is a new technology in the energy field supported greatly by the Chinese government. Since the initiation of China's "863 Program" in 1987, the Ministry of Science and Technology has organized the research and development of the key materials and technologies for NiMH and lithium-ion batteries. These batteries are produced on a large scale, particularly for export.
- In its new five-year plan (2006–2010), the Chinese government outlines steps to boost efficiency and reduce pollution. A number of clear targets for increasing energy efficiency are set (e.g., to increase total energy efficiency by 20% and to achieve an energy mix of at least 20% renewable energy by 2020). The Chinese government is also introducing clear policy standards:
    1. Starting April 1, 2006, buyers of new, large cars paid 20% more sales tax than those of new, small cars, and buyers of even smaller cars paid 1% less in sales tax.
    2. In 2006, parking fees in central Beijing were doubled.
    3. The gasoline price for consumers increased by about 20% during 2006.
    4. In December 2005, tighter vehicle standards were approved.

On the whole, the PRC is making vast progress in manufacturing lithium-ion battery technology. The government has a national program in place to attract foreign companies to set up joint ventures and/or partnerships with Chinese companies. The Chinese government offers large incentives to Chinese companies that produce batteries for export. The Chinese government also gives Chinese-owned companies additional incentives to conduct research and provides capital for manufacturing lithium-ion batteries for all applications.

1. INTRODUCTION

Argonne National Laboratory prepared this report, under the sponsorship of the Office of Vehicle Technologies (OVT) of the U.S. Department of Energy's (DOE's) Office of Energy Efficiency and Renewable Energy, for the Vehicles Technologies Team. The information in the report is based on the author's visit to Beijing; Tianjin; and Shanghai, China, to meet with representatives from several organizations (listed in Appendix A) developing and manufacturing lithium-ion battery technology for cell phones and electronics, electric bikes, and electric and hybrid vehicle applications. The purpose of the visit was to assess the status of lithium-ion battery technology in China and to determine if lithium-ion batteries produced in China are available for benchmarking in the United States. With benchmarking, DOE and the U.S. battery development industry would be able to understand the status of the battery technology, which would enable the industry to formulate a long-term research and development program. This report also describes the state of lithium-ion battery technology in the United States, provides information on joint ventures, and includes information on government incentives and policies in the Peoples Republic of China (PRC).

1.1. Scope of the Study and Objectives

The scope of this study is (1) to determine the state of the art and current production of lithium-ion batteries in China and (2) to develop recommendations for DOE with respect to battery benchmarking and testing of candidate batteries for use in hybrid and plug-in hybrid vehicles. China is the largest country in the world, and its economy is growing rapidly. Because of the demand for world oil supplies, the United States is interested in the capabilities of Chinese manufacturers of motor vehicles to produce and use state-of-the-art, energy-efficient vehicles. Although there are significant issues of competitive concern, there are also reasons to hope that multiple nations will have the ability to produce high-quality, interchangeable battery packs for future plug-in hybrid vehicles. The Chinese government is developing its industry and universities to carry out the research and development (R&D) in lithium-ion battery technology for portable and electric vehicle applications. An estimated 400 organizations in China are involved in battery development or manufacturing; however, manufacturers of lithium-ion batteries represent an unknown fraction of this total.

Some U.S. and Chinese companies have begun to develop joint relationships to conduct R&D and manufacturing of advanced vehicle technologies. The study described in this report on Chinese lithium-ion battery technology could help U.S. companies focus on technologies for which the United States has a competitive advantage. By properly selecting the best Chinese batteries for possible follow-up benchmarking and testing, this study has the potential to strengthen both U.S. and Chinese programs by avoiding undesirable duplication of effort, helping both nations to focus on their strengths with respect to the manufacturing and development of lithium-ion batteries.

Further, DOE's OVT Program always strives to determine the best technologies available for potential use by U.S. automobile manufactures and to build the capability of domestic manufacturers to produce competitive or superior technologies. Similarly, the Energy Storage effort within OVT — through battery testing and benchmarking of Japanese batteries — has also provided U.S. manufacturers and customers with a degree of assurance about the reliability of battery technologies. In the process, the Energy Storage effort has helped provide domestic manufacturers information needed to support internal decisions to proceed with programs to develop hybrid and plug-in hybrid vehicles. Greater confidence in the ultimate marketability of hybrid powertrain technologies will help ensure the success of the vehicle development program.

The purpose of this study is consistent with past efforts of the U.S. Advanced Battery Consortia (USABC) and FreedomCAR Partnership for the development of battery technologies — which will benefit from the results of this study. It provides a lithium-ion battery technology assessment specific to China, as well as information on contacts with key organizations and manufacturers. As DOE embarks on a new "Plug-In Hybrid Initiative," this study identifies lithium-ion battery technologies compatible with powertrain technologies being developed in the United States, increasing the likelihood that such powertrains can help reduce the rate of growth in oil consumption.

1.2. Methodology and Approach

Five battery research and testing organizations, one battery technology society, and five companies conducting research and development of lithium-ion battery technology (as well as manufacturing lithium-ion batteries) were visited in the PRC. During this visit, 28 interviews were conducted (Appendix A). A presentation was made to each of these organizations on the status of lithium-ion battery technology for electric, hybrid, and plug-in hybrid applications in the United States (see Appendix B). The interviewed individuals included those from industry, government laboratories, and academia. Individuals interviewed from industry included representatives from materials suppliers and representatives from battery manufacturers serving in technology development, management, and marketing positions. Each interviewee received a list of questions in advance that served as a guide to the interview process. (Appendix C lists the questions used to guide the personal interviews.) Interviews did not always follow the sequence of the listed questions. The interviews were conducted in a relaxed setting and in a conversational manner, which helped the experts to focus on what they considered to be the most important factors influencing the development of lithium-ion battery technology and the decisions of manufacturers about the production of lithium-ion batteries. Responses from those interviewed helped Argonne identify and analyze developments in lithium-ion battery technology in the PRC and to obtain information about cost estimates and manufacturing capabilities.

The initial contacts in the United States were made by attending advanced battery technologies meetings, seminars, and conferences, such as the Advanced Automotive Battery Conference 2007. The contacts and conferences provided information on developments in lithium-ion battery technology in the United States, Europe, and Asia. These conferences were extremely useful in understanding the status of lithium-ion batteries, in comparison with activities related to lithium-ion batteries in the PRC.

1.2.1. Overview and Discussion

Lithium-ion batteries offer higher power and energy per unit weight and volume, as well as a better charge efficiency, than NiMH batteries. These attributes allowed them to capture a major part of the portable rechargeable battery market within only a few years of their introduction and to generate global sales estimated at $5 billion in 2006. Nevertheless, the tolerance of lithium-ion technology to abuse is still questionable, its reliability for automotive applications is not proven, and currently, its cost is higher than that of NiMH. Over the last five years, most automakers have started to evaluate the suitability of lithium-ion batteries for hybrid electric vehicle (HEV) applications, and some have even embarked on significant in-house lithium-ion battery development projects. In the future, the lithium-ion battery is likely to become the battery of choice for most hybrid applications. When exactly this will happen for each of the hybrid vehicles is a question that automakers, battery developers, their supply chains, and the many industry stakeholders and observers are struggling to answer.

1.2.2. Lithium-Ion Battery Technology — Technology Discussion

Mass-manufacturers of lithium-ion cells for consumer products are now engaged in the development of lithium-ion chemistries for HEV and plug-in hybrid electric vehicle (PHEV) applications, with possible commercialization as early as 2012. The major impediment to developing lithium-ion batteries for PHEVs appears to be that the requirements for PHEV batteries are insufficiently defined at this time. The apparent interest in PHEVs by General Motors Corporation (GM) might stimulate efforts to develop lithium-ion technology for PHEV applications. Several companies in Europe and Japan have been developing medium- and high-energy lithium-ion technologies, some of them based on advanced materials, chemistries, and/or manufacturing techniques. Their strategy is to pursue limited-volume applications and markets that may be emerging, especially for small battery-powered electric vehicles (EVs); electric bikes; and, more recently, PHEVs. Several of these companies hold the view that lithium-ionpowered PHEVs and small battery-powered electric vehicles will be able to match the life-cycle cost-competitiveness of conventional vehicles in urban fleet applications, and a few have established cell-production capacities for hundreds to a few thousand of 10–25-kWh batteries per year, which may be sufficient for demonstration fleets.

As the lightest metal and most electronegative element, lithium is the most attractive negative electrode material for high-energy batteries. However, its high reactivity with water and with the solvents used in organic battery electrolytes has prevented its use in rechargeable batteries until two important discoveries were made about 15 years ago: lithium can be inserted ("intercalated") electrochemically in carbon "host" materials, and a protective layer forms at the interface of the lithium-containing carbon with the organic electrolyte solvent when a cell is charged for the first time. Remarkably, this complex solid-electrolyte interface (SEI) layer prevents further attack of the electrolyte by lithium but allows the passage of lithium ions during charge-discharge cycling. The host material forming the negative electrode in lithium-ion cells is made from special grades of graphite and/or coke. Mixed with binders, these carbons are deposited on thin copper sheets that serve as conducting supports. A variety of materials can be paired with carbon-based negatives in battery cells by using organic electrolytes. Mixed with carbon for increased conductivity and with binders, these materials are deposited on thin aluminum sheets as conducting supports. Currently established cathode electrode materials are listed in Table 1 and are reviewed below in the context of current lithium-ion battery technology.

LiCoO$_2$: Lithiated cobalt oxide is the main component of the positive electrodes in lithium-ion cells produced on a very large scale for consumer product applications. It has good storage capacity for lithium ions, adequate chemical stability, and good electrochemical reversibility. However, it is relatively more expensive per kilowatt-hour of storage capacity than other oxides and is therefore not a good candidate for automotive applications of lithium-ion batteries that are under severe cost constraints.

Li(Ni$_{0.85}$Co$_{0.1}$Al$_{0.05}$)O$_2$: Commonly termed NCA, this lithiated mixed oxide of nickel, cobalt, and aluminum has become accepted for batteries in prototypical HEV, full performance battery electric vehicle, and the PHEV. It approaches the favorable characteristics of LiCoO$_2$ at a lower per-kilowatt-hour cost.

Li(Ni$_{1/3}$Co$_{1/3}$Mn$_{1/3}$)O$_2$: Often termed NCM, this lithiated mixed oxide of nickel, cobalt, and manganese is potentially less expensive than NCA. It can be charged to two cell voltage levels. At the higher voltage (e.g., 4.1–4.2 V), NCM yields excellent storage capacity and relatively low per-kilowatt-hour cost but tends to degrade through the dissolution of manganese; at lower voltage, its capacity is substantially less and the per-kilowatt-hour cost is higher, but stability appears adequate.

LiMnO$_2$: Lithium manganese spinel, denoted LMS, is more stable than cobalt oxide and nickel oxide-based positives in lithium-ion cells because the spinel crystal structure is inherently more stable and has no or little excess lithium ions in the fully-charged state. Thus, it provides very little lithium for undesirable lithium metal deposition on the negative electrode in overcharge. Also, the threshold of thermal decomposition of the charged (lithium-depleted) material is at a considerably higher temperature than that of other positive electrode materials. Despite its lower specific capacity, the expected substantially lower per-kilowatt-hour cost will make LMS attractive, if the efforts to stabilize the material against electrochemical dissolution of its manganese content are successful.

LiFePO$_4$: Lithiated iron phosphate (olivine), denoted LFP, is now being used successfully as a potentially lower-cost positive electrode material. Because of its lower electrochemical potential, LFP is less likely to oxidize the electrolyte solvent and thus is more stable, especially at elevated temperatures. Doping is used to increase the conductivity and stability of this promising material.

Table 1. Cathode Electrode Materials

Active Material Chemical Formula (discharged state)	Storage Capacity (mA·h/g)	Normal Voltage (V)	Wh/kg	Wh/L	Material Cost Range ($/kg)	Material Cost Range ($/kWh)
LiCoO$_2$	145	4.0	602	3,073	30–40	57–75
Li(Ni$_{0.85}$Co$_{0.1}$Al$_{0.05}$)O$_2$	160	3.8	742	3,784	28–30	50–55
Li(Ni$_{1/3}$Co$_{1/3}$Mn$_{1/3}$)O$_2$	120	3.85	588	2,912	22–25	30–55
LiMnO$_2$	100	4.05	480	2,065	8–10	20–25
LiFePO$_4$	150	3.34	549	1,976	16–20	25–35

Notes:
- Lower potential can provide greater stability in electrolyte
- Cobalt oxide is most widely used in consumer cells but recently too expensive; LiMn$_{1/3}$Co$_{1/3}$Ni$_{1/3}$O$_2$ newer than LiNiCoO$_2$
- Mn$_2$O$_4$ around for many years – not competitive for consumer – good for high power
- LiFePO$_4$ – very new – energy density too low for consumer electronics – safe on overcharge; however, need electronics to prevent low voltage

The electrolyte used in lithium-ion battery cells is a solution of a fluorinated lithium salt (typically $LiPF_6$) in an organic solvent, enabling current transport by lithium ions. Separators are usually microporous membranes made of polyethylene or polypropylene. Because of the low conductivity of organic electrolytes, adequate cell and battery power can be realized only with electrodes and separators that are much thinner than those used in aqueous-electrolyte batteries. The need for thin electrodes has made spiral winding of positive electrode-separator negative electrode composites the preferred method for the fabrication of lithium-ion cells, but flat cell configurations packaged in soft plastic (often metallized) enclosures are now gaining acceptance.

1.3. Lithium-Ion Battery Thermal Management

Battery thermal management is critical in achieving adequate performance and extending life of batteries in electric and hybrid vehicles under real driving conditions. Designing a battery thermal management system for given HEV/PHEV battery specifications starts with answering a sequence of questions:

- How much heat must be removed from a pack or a cell?
- What are the allowable temperature maximum and difference?
- What kind of heat transfer fluid is needed?
- Is active cooling required?
- How much would the added cost be for the system?

To find a high-performance and cost-effective cooling system, it is necessary to evaluate the system thermal response and its sensitivity as a function of controllable system parameters.

Battery temperature affects battery performance and life. Therefore, battery thermal management is a critical element for achieving the desired performance and calendar life for battery packs in HEVs. Automakers and their suppliers are paying increased attention to battery thermal management to ensure that warranty costs for battery replacement are reasonable. The battery in an HEV experiences a demanding thermal environment and may need to be heated during cold-weather operation and cooled during extended use and during warm-weather operation. A uniform temperature should be maintained among the battery's cells because cellto-cell temperature variability leads to imbalances and reduced performance and it potentially reduces calendar life.

In the thermal design process, researchers should consider the cell-to-cell variability in a multi-cell pack, which could lead to different battery electrical and thermal behavior. There is also variability in the mechanical design and method for heating or cooling each cell. Also during the thermal design process, researchers should consider the impact of various design parameters, such as state of charge (SOC), internal resistance, current amplitude, heat-generation rate, fluid flow rate, cooling/heating fluid temperature, and various geometrical variations. The goals of a battery thermal management system are to keep the battery below a certain temperature and to minimize the temperature variation in the pack while using a minimum amount of energy.

1.4. Air Cooling versus Liquid Cooling

The choice of a heat-transfer medium has a significant impact on the performance and cost of the battery thermal-management system. The heat-transfer medium could be air, liquid, a phase-change material, or any combination of these media. Heat transfer with air is achieved by directing/blowing the air across the modules. However, heat transfer with liquid could be achieved by using discrete tubing around each module, using a jacket around the module, submerging modules in a dielectric fluid for direct contact, or placing the modules on a liquid heated/cooled plate (heat sink). If the liquid is not in direct contact with modules, such as in tubes or jackets, the heat-transfer medium could be water/glycol or even refrigerants, which are common automotive fluids. If modules are submerged in the heat-transfer liquid, the liquid must be dielectric, such as silicon-based or mineral oils, to avoid any electrical shorts.

Using the air as the heat-transfer medium may be the simplest approach, but it may not be as effective as heat transfer by liquid. The rate of heat transfer between the walls of the module and the heat-transfer fluid depends on the thermal conductivity, viscosity, density, and velocity of the fluid. For the same flow rate, the heat-transfer rate for most practical direct-contact liquids (such as oil) is much higher than that with air because of a thinner boundary layer and a higher fluid thermal conductivity. However, because of oil's higher viscosity and associated higher pumping power, a lower flow rate is usually used, making the heat-transfer coefficient of oil only 1.5–3 times higher than that of air. Indirect-contact heat-transfer liquids (such as water or water/glycol solutions) generally have a lower viscosity and a higher thermal conductivity than most oils, resulting in higher heat-transfer coefficients. However, because the heat must be conducted through walls of the jacket/container or fins, the effectiveness of indirect contact decreases.

1.5. Active versus Passive Systems

Because of cost, mass, and space considerations and their use in mild climates, battery packs in early vehicles — particularly EVs — did not use heating or cooling units and depended on the blowing of ambient air for the rejection of heat from the batteries. Early prototype HEVs also used passive ambient air-cooling. Current production HEVs (Honda Insight and Toyota Prius) use cabin air for cooling/heating of the pack. Although the ambient air is heated and cooled by the vehicle's air-conditioning or heating system, it is still considered to be a passive system. For passive systems, the ambient air must have a mild temperature (10–35°C) for the thermal management to work; otherwise, the performance of the pack can suffer in very cold or very hot conditions. Outside of these conditions, active components (such as evaporators, heating cores, engine coolant, or even electric and fuel-fired heaters) are needed.

1.6. Cooling-Only Systems versus Cooling and Heating Systems

Electric vehicles were initially aimed for the mild to warm climate of California. Battery performance is generally better at higher temperature; however, battery life can decrease with higher temperature. Therefore, batteries in those EVs needed to be cooled only, and there was no need for too much heating. At cold temperatures (below −10°C), the energy/power capability of most batteries diminishes, and electric vehicle (EV) and HEV performance diminishes as well. Heating systems have been used for EVs operating in colder climates. For EVs, there is no engine to aid in heating the battery pack, so the heat rejected from motor and power electronics and electricity from the battery could be used for heating; otherwise, a fuel-fired heater could be considered. For HEVs, the heat from the engine could be used, but it would take some time (more than 5 min) for the engine to start warming the batteries. Because power from the battery is needed much sooner, self-heating battery technology could be an option. Cooling the batteries is a less-challenging task than heating because the vehicle's air-conditioning/refrigerant system or engine coolant could be used. Energy use, however, increases with the use of refrigeration, which is contrary to the HEV goal of improving fuel economy.

1.7. Series versus Parallel Air Distribution

There are two methods for distributing air to a pack for cooling and/or heating. The first method is *series* cooling, in which air enters from one end of the pack and leaves from the other, exposing the same amount of air to several modules. The second method is *parallel* cooling, in which the same total airflow is split into equal portions, and each portion flows over a single module. Depending on the size and geometry of the modules, a series-parallel combination could be configured. Generally, *parallel* airflow provides a more even temperature distribution among the modules in a pack. The packs in GM's EV1, Toyota's RAV4-EV, Honda's Insight HEV, and Toyota's Prius (Japanese version) all have either series or series-parallel air distribution. The Toyota Prius (North American version) uses a pure parallel air distribution system or even- temperature distribution. In parallel flow design, distributing airflow uniformly to a large battery pack requires a careful design of the air manifold.

1.8. Thermal Management for Valve-Regulated Lead-Acid, NiMH, and Lithium-Ion Batteries

The relative need for the thermal management of each of the valve-regulated lead-acid (VRLA), nickel metal hydride (NiMH), and lithium-ion batteries depends on the heat-generation rate from each type of battery, its energy efficiency, and the sensitivity of performance to temperature. From Table 2, it can be seen that NiMH batteries generate the most heat at high temperatures (>40°C) and are the least efficient. At room temperature, NiMH generates less heat than VRLA and lithium-ion batteries. The performance of a NiMH battery is more sensitive to temperature than VRLA and lithium-ion batteries. Therefore, NiMH batteries need a more involved battery management control. This is also evident from various efforts to use the more effective liquid cooling for NiMH batteries. The concerns for lithium-ion packs are safety and their relatively poor performance at very cold temperatures. Because lithium-ion batteries can deliver much more power and thus more heat for the same volume than either VRLA or NiMH, heat removal must be efficient. Thermal management also depends on the type of vehicle and the location of the pack. For EVs and series HEVs, the pack is generally large, and its thermal management system may need to be more elaborate — possibly incorporating liquid cooling (particularly for NiMH). However, for parallel HEVs, the pack is generally smaller, and the thermal control could be achieved by a simpler air cooling/heating design, especially for lithium- ion and VRLA batteries.

A well-designed thermal management system is required to regulate EV and HEV battery pack temperatures evenly, keeping them within the desired operating range. Production HEVs require an active heating/cooling battery temperature monitoring system to allow them to operate in hot and cold climates. Proper thermal design of a module has a positive impact on the overall management and behavior of the pack. A thermal management system using air as the heat- transfer medium is less complicated, although also less effective, than a system using liquid cooling/heating. Generally, for parallel HEVs, an air thermal management system is adequate, whereas for EVs and series HEVs, liquid-based systems may be required for optimum thermal performance. The NiMH batteries require a more elaborate thermal management system than lithium-ion and VRLA batteries. Lithium-ion batteries also need a good thermal management system because of safety and concerns about low-temperature performance. The location of the battery pack may also have a strong impact on the type of battery thermal management and whether the pack should be air-cooled or liquid-cooled.

1.9. Heat Generation and Heat Capacity

The magnitude of the overall heat-generation rate from a battery pack under load dictates the size and design of the cooling system. The heat generation (due to electrochemical enthalpy change and electrical resistive heating) depends on the chemistry type, construction, temperature, state of charge, and charge/discharge profile. At the National Renewable Energy Laboratory (NREL), a custom-built calorimeter to measure the heat generation from cells/modules with various cycles, states of charge, and temperatures is used. Table 2 shows some typical results for various batteries from an experiment by NREL. These and other data show that, for the same current draw, a NiMH battery generates more heat than VRLA or lithium-ion batteries at elevated temperatures (>40°C). Heat generation from VRLA and lithium-ion batteries is roughly the same for similar currents. At room temperature, less heat is generated for NiMH for the same current, but NiMH is not as energy-efficient. Generally, as temperature decreases, more heat is generated because of an increase in resistance in the cells. As the discharge rate increases, more heat is generated. Under certain conditions, the battery electrochemical reaction could be endothermic, as shown in Table 2 for lithium-ion batteries at a C/1 discharge rate at 50°C.

1.10. Lithium-Based Batteries: Advantages and Challenges

A key advantage of lithium-based batteries is the high cell voltage, the direct result of the highly negative potential of lithium. With currently used mixed-oxide positives, the lithium-ion cell operating voltage range is approximately 2.75–4.2 V. The nominal (average) discharge voltage is about 3.6 V, and most of the usable cell capacity is delivered between 4.0 and 3.5 V. With iron phosphate positives, the nominal cell voltage is about 3.4 V. The high cell voltage is the fundamental reason for the high specific energy of lithium-ion cells and batteries. The high cell voltage also results in a smaller number of cells for a battery of given voltage, for reduced fabrication costs and increased reliability. A second basic advantage of the lithium-ion electrochemistry is based on the small size of lithium, which permits reversible electrochemical intercalation of lithium atoms into carbon-based negative electrodes with little structural stress and strain. Similarly, the very small lithium ion is readily and reversibly incorporated into a variety of host oxides that form the positive electrode. These characteristics are responsible for maintaining the integrity of both electrodes during charge-discharge cycling, a key requirement for long cycle life — especially in deep-discharge cycling. As discussed further below, key technology advantages of lithium-ion batteries are high power density and energy efficiency as a result of thin-cell construction and low self-discharge rate.

The main challenges encountered in the development of lithium-ion technologies for practical applications are also due to the highly negative potential of lithium. It is a powerful driving force not only for the effectiveness of lithium as a negative electrode but also for its chemical reactivity within the cell. Only the formation of an SEI prevents continued, uncontrolled reaction of lithium with the electrolyte solvent and enables the controlled discharge and recharge of lithium-ion cells and batteries. Once formed as a protective thin layer, the SEI must be stabilized chemically and kept from growing thicker because of the associated irreversible declines of cell capacity (through the loss of lithium) and of peak power (through growth of cell resistance). Choosing proper electrolyte solvents and additives and keeping cell temperatures below approximately 45–50°C are very important for stabilizing the SEI and achieving practical calendar and cycle life.

Another key challenge is the sensitivity of lithium-ion cells to overcharge that can result in chemical decomposition of positive electrode materials and the electrolyte and/or in the deposition of metallic lithium at the negative electrode. These processes damage the cell and can result in hazardous conditions, including gassing and release of flammable electrolyte solvent vapors, if the cell safety seal is breached as a result of excessive gas pressure. To avoid overcharge, lithium-ion batteries require accurate voltage control for every cell, unlike NiMH and other aqueous electrolyte batteries that can tolerate significant amounts and rates of overcharge. Accurate and reliable control of cell voltage and temperature is thus critical requirements for achieving long life and adequate safety of lithium-ion batteries for all uses, but especially so for automotive applications, which demand a very long battery life and high levels of safety.

Table 2. Heat Generation from Typical HEV/EV Modules Using NREL's Calorimeter

Battery Type	Cycle	Heat Generation W/Cell		
		0°C	22–25°C	40–50°C
VRLA, 16.5 A•h	C/1 Discharge, 100% to 0% SoC[a]	1.21	1.28	0.4
VRLA, 16.5 A•h	5C Discharge, 100% to 0% SoC	16.07	14.02	11.17
NiMH, 20 A•h	C/1 Discharge, 70% to 35% SoC	-	1.19	1.11
NiMH, 20 A•h	5C Discharge, 70% to 35% SoC	-	22.79	25.27
Lithium-Ion, 6 A•h	C/1 Discharge, 80% to 50% SoC	0.6	0.04	-0.18
Lithium-Ion, 6 A•h	5C Discharge, 80% to 50% SoC	12.07	3.50	1.22

[a] SOC = state of charge.

2. LITHIUM-ION BATTERIES: STATE OF THE ART

2.1. Performance and Life

For more than a decade, prospective manufacturers have been developing lithium-ion batteries for electric vehicles. A number of these efforts were terminated when the initiatives to introduce electric vehicles were abandoned earlier in this decade. However, some programs continued and resulted in the development of high- or medium-energy/medium-power lithium- ion technologies. Although none of the programs have generated commercially available batteries as yet, a few have resulted in the low-volume production of cells and in-vehicle evaluation of prototype batteries. Key characteristics of these technologies are summarized in Table 3.

Table 3 indicates that current designs of high-energy cells achieve energy and power density levels of at least 150 Wh/kg and 650 W/kg. Batteries of 20–30 kWh using such cells can attain energy densities of around 100 kWh/kg and power densities of 250–3 50 W/kg or above, which is sufficient for small or even full-performance electric applications at acceptable battery weights. Also, medium-power lithium-ion cells in the appropriate size range enable construction of 7.5–15-kWh batteries with energy densities above 70 Wh/kg and power densities in the 5 00–900 W/kg range, values that can readily meet the performance requirements of PHEV batteries.

Table 3. Status of Lithium-Ion High-Energy/Medium-Power Cell and Battery Technologies

Component	Manufacturer				
	JCS	GAIA	LitCEL	Lamilion	Kokam
Cell	VL 45E/41M	HE/HP Series	EV Type	EV Type	HE/HP
Voltage (V)	3.6	3.6	3.85	3.6	3.7
Capacity (A•h)	45/41	60/45	50/33	13	100/40
Energy Density (Wh/kg)	150/136	150/105	136/142	~150	163/135
Energy Density (Wh/L)	314/286	3 80/284	270	270	340/285
Peak Power Density (W/kg)	664/794	900/1,500	1,500	1,300	~700/~1250
Power/Energy Ratio (L/h)	4.4/5.8	~6/~14	7.7	8.7	~4.3/~9
Cycle Life (Cyc. @% DOD)	~3,200 (80)	~1,000 (70)	~1,000	~1,400 (100)	~3,000 (80)
Calendar Life (years at RT)	~12	–	–	~10	~10
Development Status	LP	LP	LP	LP	LP
Battery (Applications)	EV/PHEV	EV/PHEV	EV/PHEV	Small EV	EV/PHEV
Storage Capacity (kWh)	~24/15	22/8.1	20/7.6	9.2	~30/~5
Energy Density (Wh/kg)	90/94	115/74	118/117	~70	~1 10/~100
Energy Density (Wh/L)	145/80	165/130	194	–	–
Peak Power (kW)	55/87	50/80	155/60	62	13 0/47
Peak Power Density (W/kg)	210/540	~250/730	912/917	~400	~490/~940
Power/Energy Ratio (1/h)	2.3/4.6~2.2/~10	7.7/7.8	6.7	~4.3/~9.4	
Weight (kg)	265/160	200/110	170/65	150	265/~50
Development Status	LP;VE	LP;VE	LP; VE	LP; VE	LP

Table 3 also includes data on cycle and calendar lives, which are two of the remaining concerns about the readiness of lithium-ion batteries for vehicle applications. The more than 3,000 deep cycles achieved by Saft (a "world specialist in the design and manufacture of high-tech batteries for industry"), and also claimed by Kokam, indicate that large lithium-ion cells have the potential for very long cycle life. The calendar life of state-of-the-art, high-energy lithium-ion technology is also much improved over the values of 2–4 years that were typical five years ago.

Table 4 presents performance and life data for high-power/medium-energy lithium-ion cell and battery technologies for HEV applications. This table shows a promising lithium-ion high-power technology in a smaller cell size that has been commercialized by A1 23 Systems for power tool applications. The basic technology is expected to be developed into larger cell sizes of higher power density for HEVs, an application for which it is well suited because of the safety advantages of iron phosphate-based positives.

Even smaller lithium-ion cells of the type used in consumer electronic products are being used for developmental PHEVs and full performance battery electric vehicles (FPBEVs) in the form of batteries that consist of several thousand cells connected in parallel and in series. This approach takes advantage of lithium-ion cells that are available now, since they are being produced in very large numbers and sold at competitive prices for laptop computers. However, it raises questions regarding the reliability, safety, and ultimately achievable cost of "small-cell" batteries.

Table 4. Status of Lithium-Ion High-Power/Medium-Energy Cell and Battery Technologies

Component	Manufacturer					
	JCS	Matsushita	HitachiVE	Kokam	GAIA	A123Systems
Cell	VL7P	Gen 2	UHP	HP	MI 26650	–
Voltage (V)	3.6	3.6	3.4	3.7	3.6	3.3
Capacity (A·h)	7	7	5.5	7.2	7.5	2.3
Energy Density (Wh/kg)	67	92	–	114	84	110
Power Density (W/kg)	1,800	3,400	–	2,600	1,500	1,950
Power/Energy Ratio (1/h)	27	37	–	23	18	20
Power Density (W/L)	3,525	–	–	4,900	3,750	4,200
Cycle Life (shallow cycles)	~400k	–	~750k	–	–	~240k
Cycle Life (cycles/DOD)	–	~1,000	–	~3,000/80	1,000/60	7,000/100
Calendar Life (years@RT)	~20	–	–	~10	–	~15
Development Status	LP	–	–	LP	LP	CP
Battery (Application)	HEV	HEV	HEV	HEV	HEV	PT
Storage Capacity (kWh)	2	3	1	2.6	2	–
Peak Power (kW)	50	90	47	52	25	2.1
Peak Power Dens. (W/kg)	1,110	2,100	1,900	1,850	–	–
Power Density (W/L)	1,110	--	2,100	–	–	2,200
Energy Density (Wh/kg)	44	70	42	93	–	–
Power/Energy Ratio (1/h)	25	30	45	20	12.5	–
Weight (kg)	45	43	22.5	28	–	–
Development Status	PP	D	PP	LP		

The calendar life of high-power lithium-ion battery cells is expected to depend on temperature much in the same way as high-energy cell designs, because several of the high-power cell technologies use the same basic chemistry as larger cells and thus are subject to the similar degradation processes.

Table 5 provides information on lithium-ion battery developers and on types of cathodes, anodes, packaging, cell structures, and cell shapes used in battery developments.

As with high-energy lithium-ion technologies for battery-powered electric vehicles and PHEV applications, a key technical issue is whether developers can meet the very high levels of safety required for vehicles operated on public roads.

2.2. Cost

The current cost of lithium-ion HEV batteries made with early pilot equipment is around $2,000/kWh. If the current designs were brought to volume production, it is anticipated that the cost would drop to approximately $1,000/kWh.

The cost calculation procedure for a lithium-ion battery module using 10-A·h $LiMn_2O_4$ HEV cells is illustrated in the following discussion. The cell's material cost is estimated at $9.93, with the key cost drivers classified by order of importance as follows: (1) separator, (2) electrolyte, (3) $LiMn_2O_4$, (4) graphite, and (5) copper foil — these five factors account for 75% of the cell's material cost.

Assuming that materials before yield losses represent 65% of the cost of goods (considerably less than that for NiMH because of the lower yield and higher depreciation for lithium-ion technology), and with a low gross margin of 30%, the calculated cost of a cell is $606/kWh. Further, assuming that the cost of the module per kilowatt-hour is 1.5 times the cost of the cell, and that the pack cost per kilowatt-hour is 1.43 times that of the module, we arrive at a module cost of $1,011 and a pack cost of $1,444/kWh.

The following analysis of the "best-case" scenario is more uncertain than that for the NiMH battery, because:

- The cathode material, $LiMn_2O_4$, is not yet in high-volume production.
- The separator and the electrolyte are the top two items driving the cost of materials. In both cases, the high cost is a result of the difficulty of making these high-purity (electrolyte) and high-dimensional-accuracy (separator) materials, even though the cost of the underlying raw materials is quite moderate.
- Yield is a significant item affecting cost in early years of production; the allowable factor for the yield of materials is attributed as 65%, as compared with 75% in the case of NiMH.
- The cost of electrical management for the module and pack is very high at present. Although reductions in cost will occur, they are difficult to estimate at this point. Under the "best-case" scenario, there will be significant reductions in the cost of materials and in module and pack peripheral costs. Certain advances in materials — possibly including new electrolyte salts, new cathode material, and a less-expensive process to make the separator — may occur. Such innovations, when fully developed and tested, could lower the price of the battery. This may take 5 to 8 years.

Table 5. Developers of Lithium-Ion Technology Cells for HEV Applications

Company	Cathode	Anode	Packaging	Structure	Shape
Toyota	NCA	graphite	metal	spiral	prismatic
Panasonic	NMC	blend	metal	spiral	prismatic
JCS	NCA	graphite	metal	spiral	cylindrical
Hitachi	NMC/LMO	hard carbon	metal	spiral	cylindrical
NEC-Lamilin	LMO/NCA	hard carbon	pouch	stacked	prismatic
Sanyo	NMC/LMO	blend	metal	spiral	cylindrical
GS Yuasa	LMO/NCA	hard carbon	metal	spiral	prismatic
A123 Systems	LFPO	graphite	metal	spiral	cylindrical
LG Chem.	LMO	hard carbon	pouch	stacked	prismatic
Samsung	LMo/NMC	graphite	metal	spiral	cylindrical
SK Corp.	LMO	graphite	pouch	spiral	prismatic
EnerDel	LMO	LTO	pouch	spiral	prismatic
AltairNano	NMC/LCO	LTO	pouch	stacked	prismatic

In 1994, the cost of 18650 type (rated at 1,1 00-mA•h) cylindrical consumer lithium-ion cells was over $10 per cell at a high volume. In 2001, the capacity of the cell increased to 1,900-mA•h, and the cost dropped to around $2.00 per cell at the comparable volume, resulting in a drop in cost of $3/Wh for a comparable 1,1 00-mA•h cell. This is a remarkable improvement in less than seven years. The recent price of this cell — now made in quantities of over 100 million per year — is about $300/kWh. The HEV cell, with its ultra-thin electrodes, uses approximately twice the amount of separator and electrolyte per watt-hour as the consumer cell; this has a significant impact on the cost of materials and manufacturing, which is determined to a large extent by the design of the electrodes.

2.3. Lithium-Ion Battery Safety

Battery safety is critical for the success of lithium-ion batteries for HEV, PHEV, and EV applications. Lithium-ion battery safety is tied directly to the avoidance or strict control of those processes in lithium-ion battery cells that, if uncontrolled, can release dangerous amounts of energy, flammable gases, and/or toxic chemicals into the battery environment. These processes include (1) electrochemical overcharging of battery cells and the ensuing reactions of the chemical species formed during overcharge and (2) chemical reactions of the organic electrolyte/solvent with one or both electrodes. Under normal operating conditions of cell voltage and temperature, these processes are either precluded through cell-level voltage control (overcharge) or occur at very low rates that do not constitute safety risks.

Concerns about lithium-ion battery safety thus can be limited to the response of cells and batteries to "abuse," including electrical/electrochemical (shorting, high rate, and extensive overcharging), thermal (heating to temperatures above the cell tolerance limit), and/or mechanical (destruction of physical integrity). Abuse tolerance testing has become part of cell development, as well as battery design and engineering efforts. The degree of tolerance to various abuses is serving as a relative measure of safety and as a guide to the development of adequately safe lithium-ion cells and batteries. The procedures most commonly used in lithium- ion abuse testing were developed with DOE funding under the USABC and FreedomCAR programs and are now widely accepted. Results of systematic abuse testing of small commercial lithium-ion cells following these procedures show that sustained high-rate/high-voltage overcharge and massive shorting of some lithium-ion cell types can cause thermal runaway that is accompanied by cell-internal gas evolution, cell venting, and (if triggered by sparks) burning of vented electrolyte solvent. However, these conditions can be created only if the standard, multiple levels of protection devices (e.g., voltage-sensitive and pressure-driven switches to interrupt current; current-sensitive and temperature-activated fuses; and cell-balancing electronics) are removed.

Although the chemistries of the cells and batteries used in EV and HEV battery technology are similar to the chemistries used in lithium-ion laptop batteries, the cell and battery designs are substantially different. Even more important, batteries for HEV, PHEV, and battery- powered electric vehicle applications have voltage, pressure, and temperature sensors integrated in multiple, independent controls that prevent or terminate unsafe battery conditions of the type that have resulted in some laptop battery fires.

Experience with more than 200 electric and hybrid vehicles equipped with lithium-ion batteries designed for FPBEVs or HEVs and road tested in California, Europe, and Japan over the past five years validates the high level of safety achieved for current lithium-ion technology. No significant safety issues were encountered during these tests.

While lithium-ion technology representing the state of the art of several years ago has proven safe in on-road vehicle testing, R&D is continuing to further enhance battery life and safety, as part of extensive worldwide efforts to advance all aspects of lithium-ion cell and battery technology. Challenges for developing lithium-ion battery technology are costs at initial production volumes, safety, life, and manufacturing reliability.

2.4. Lithium-Ion Battery Industry in China

The PRC is the country with the largest population in the world. The PRC has been involved in battery technology developments and manufacturing for several years. The PRC exported the largest number of batteries for telecommunication, computers, cell phones, and other electronic equipment to many countries over the last 10 years. Several hundred companies, both small and large, are involved in development of lead-acid, NiMH, and lithium-ion batteries for these applications. In the past 3 to 4 years, many outside companies have been bringing advanced battery technologies to the PRC and setting up partnerships and/or joint ventures to manufacture batteries for these and other applications (such as electric bikes, EVs, and HEVs) to take advantage of low labor costs in China and incentives provided by the Chinese government. Companies in the PRC are aggressively working toward development of manufacturing processes for the battery export market.

In 2003, the annual mobile phone production capability of the Chinese telecommunication equipment manufacturing industry reached 200 million sets, with an actual annual output of 186 million sets, among which about 120 million sets were exported — that accounts for a quarter of the total global output of mobile phones. Driven by the mobile phone market, the telecommunication equipment manufacturing industry improved its share of the export market. In 2003, Chinese companies produced 334 million batteries just for mobile phones.

Taking advantage of small early markets for lithium-ion battery technology, Chinese companies were involved in developing a large number of advanced batteries in collaboration with foreign companies. Now those companies are developing lithium-ion batteries for electric bikes, EVs, HEVs, and PHEVs. From 2001 to 2004, the number of battery companies in China increased from 455 to 613; accordingly, the number of employees also increased from 140,000 in 2001 to 250,000 in 2004. The total output reached 63.416 billion Yuan ($8.1 billion) in 2004, which is an increase of 52.58% over 2001.

The sales of batteries produced by large-scale companies in the battery industry were 59.82 billion Yuan ($7.65 billion) in 2004, which was an increase of 53% compared with sales in 2003; an increase of 105% compared with sales in 2002; and an increase of 161% compared with sales in 2001. This growth is attributed to the growth of large companies. In the last four years, the debt-to-asset ratio of China's battery industry has been fluctuating between 54 and 59%.

2.5. International Standards for the Battery Industry

The International Electrical Commission (IEC) standards are the most commonly used standards by the battery industry for testing and evaluating battery technologies. The standard for testing nickel-cadmium battery is IEC 60285; the standard for testing nickel metal hydride is IEC61436. For the evaluation of lithium-ion batteries, the most commonly used standards are those developed by SANYO or Panasonic. Some companies use IEC standard 61960 to test and evaluate lithium-ion batteries. Also, the IEEE1625 standard is used to determine the improvements in the reliability of lithium-ion batteries. This standard includes the appearance of portable computers, vibrations, environment protection of the battery unit, and assembly.

2.6. International Market Competition

Along with development of the electronic product market for mobile phones, notebook computers, digital cameras, and portable video cameras, the lithium-ion battery industry is also growing substantially. In 2003, the global output of lithium-ion batteries surpassed 1.3 billion units, with a growth rate of over 60%, while total sales were more than $4 billion. In 2005, the global output of lithium-ion batteries grew again by 48%.

During 2000–2003, the lithium-ion battery industry in China grew rapidly, at an annual rate of over 140%. At present, the lithium-ion battery market in China is 32 million units per month, which is 29% of the global market share. In 2003, the global top 10 manufacturers of lithium-ion batteries held 81% of the global market share. Four Chinese manufacturers are listed in the top 10: BYD Company Limited; Shenzhen BAK Battery Co., Ltd.; Shenzhen B&K Technology Co., Ltd.; and Tianjin Lishen Battery Joint-Stock Co., Ltd. In the future, along with the expansion in production of BYD Company Limited; Shenzhen BAK Battery Co., Ltd.; Shenzhen B&K Technology Co., Ltd.; and Tianjin Lishen Battery Joint-Stock Co., Ltd., the lithium-ion battery industry in China is expected to grow at an annual average rate of more than 30%.

The Korean lithium-ion rechargeable battery industry is keeping pace with the growth of the electronics industry. In 2005, the global market share for Samsung SDI and Samsung LG

Chemistry companies' lithium-ion batteries reached 28%. By 2010, China will reportedly surpass Korea in the growth of the lithium-ion battery industry because the cost of the battery is lower than that in Korea or Japan, according to the officials at the Ministry of Science and Technology (MOST). Therefore, the future of the lithium-ion battery industry in China depends on a breakthrough in quality and performance.

The polymer lithium-ion (PLI) battery has become the leader in the Chinese battery industry, since it performs better and has a higher power density than the conventional lithium- ion battery — plus, it is also lighter and thinner. Although the price of the PLI battery is 10–30% higher than the price of the lithium-ion battery, since 1999 the rate at which the market for PLI batteries is growing is still faster than the rate at which the market for lithium-ion batteries is growing. Compared with 1999, the market share of PLI batteries increased by 8% in 2003 and by 10% in 2004. The market share of PLI batteries is increasing in China, Japan, and Korea, a trend similar to the market for lithium-ion batteries. Because of increasing market demand and expansion plans to produce PLI batteries in China, Japan, and Korea, the market share of PLI batteries will continue to increase annually.

Lithium-ion batteries are widely used in various electronic products, such as handsets, laptops, Personal Digital Assistants (PDAs), and digital video. Some electric vehicles also use lithium-ion batteries. In Japan, 57.4% of lithium-ion batteries are used in mobile phones, 31.5% in notebooks, 7.4% in digital videos, and 3.7% in other products. In China, most lithiumion batteries are used in handsets and cell phones, because laptops and digital videos have not been popularized yet. Lithium-ion batteries have become the mainstream batteries for the laptop and mobile markets.

In China, such manufacturers as BYD Company Limited; Shenzhen BAK Battery Co., Ltd.; Shenzhen B&K Technology Co., Ltd.; and TCL Corporation (and others) hold large shares of the global lithium-ion battery market. However, the capacity and quality of lithium-ion batteries need to be improved. The obstacles to improving lithium-ion battery performance include cathode development, cost, development of a functional additive, and patent protection.

2.7. Growth of Lithium-Ion Battery Technology in China

In 2003 and 2004, the lithium-ion battery industry in China was booming, and during that time, many companies established new production lines. In 2004, China had about 60 manufacturers of lithium-ion batteries. Market competition has been intense; the total market share of the top 10 manufacturers in 2003–2004 was about 85%, indicating that enterprises find it difficult to compete with large manufacturers.

Along with the rapid growth in the number of lithium-ion battery manufacturers in China, companies like BYD Company Limited; Tianjin Lishen Battery Joint-Stock Co., Ltd.; Shenzhen BAK Battery Co., Ltd.; and Shenzhen B&K Technology Co., Ltd., are also increasing their market share. In 2004, the domestic and overseas markets for lithium-ion batteries were flourishing, the export volume was 189 million units, and sales increased by 16.3%. Because of the rapid increase in domestic demand and the import of 550 million units of lithium-ion batteries, sales increased by 23.4% in 2004. In the future, along with the expanded production by some key enterprises and the continuous growth of portable products such as mobile phones and notebook computers, China's lithium-ion battery industry could still maintain the annual average growth rate of more than 30%.

2.8. Key Raw Material

The rapid growth of the lithium-ion battery industry in China drives the growth of the key raw material industry for lithium-ion batteries. Before 2001, the raw materials for lithium-ion batteries were mainly imported. However, domestic sources now supply separator material, anode material, electrolyte, and cathode material for lithium-ion batteries. At the end of 2003, Gejiu City in Yunnan Province established production lines for lithium cobalt, with an annual production capability of 800 tons. In 2004, the Chinese Academy of Geological Sciences developed a new technique for extracting lithium from salt lakes and established a demonstration-engineering project for a source of salt lake lithium in Baiyin City in Gansu Province. So far, the Chinese Academy of Geological Sciences has produced 600 tons of lithium carbonate. In 2004, key manufacturing equipment for LiNiO was developed to mass-produce it at a production capability of 500 tons per year. China has abundant nickel resources; therefore, establishing LiNiO production lines will alleviate the import situation. Chinese companies will not have to depend on importing anode material for lithium-ion batteries. In early 2004, the production of carbon material for lithium-ion battery anodes surpassed 1,000 tons.

At present, China has a relatively well-developed lithium-ion battery manufacturer base. Manufacturers such as BYD Company Limited; Shenzhen BAK Battery Co., Ltd.; and Shenzhen B&K Technology Co., Ltd., enjoy a large share of the global battery market. During 2003–2004, the Chinese lithium-ion battery industry developed dramatically. The production of cobalt acid lithium and nickel acid lithium, as well as the invention of a new manufacturing technique to extract lithium from salty lakes, could drastically reduce the importation of anode materials for lithium-ion batteries.

3. LITHIUM-ION BATTERY TECHNOLOGY IN CHINA

The demand for lithium-ion rechargeable batteries has been driven by the rapid growth in the use of electronic portable equipment, such as cellular phones, laptops, and digital cameras. In addition, the expectation that rechargeable batteries will play a large role in alternative energy technology, as well as in electric bikes (e-bikes), EVs, hybrid vehicles, and PHEVs, has made the development of lithium-ion rechargeable batteries a fast-growing industry in China and the world. The first commercial lithium-ion rechargeable battery, introduced by Sony Japan in 1989, used graphite as the anode. Since then, Chinese companies have been developing and producing lithium-ion batteries for portable applications. Recently, Chinese companies that manufacture lithium-ion batteries grew by several hundred. Large companies (such as BYD Company Limited; Shenzhen BAK Battery Co., Ltd.; Shenzhen B&K Technology Co., Ltd.; Tianjin Lishen Battery Joint-Stock Co., Ltd.; and others) hold a large share of the global market. The performance characteristics of lithium-ion cells from various manufacturers are given in Table 6.

The capacity of lithium-ion cathode materials needs to be improved. Obstacles to improving electrolyte include cost and functional additives, as well as foreign patent protection.

Lithium-ion batteries, whether for EVs, e-bikes, or consumer electronics, are all produced by using similar processes, described in depth in Gaines and Cuenca (2000). Hence, a single manufacturer can produce battery sizes for a wide range of applications, from portable consumer electronics to EVs. Lithium-ion batteries can be designed for high power or high energy, depending on cell size, thickness of the electrode, and relative quantities of the material used. High-power cells are generally smaller in order to dissipate the higher heat load. Both types use the same current collectors and separators. Lithium resources are abundant in China. As of 2000, China was the second largest producer of lithium in the world and, in 2004, it produced 18,000 metric tons.

Table 6. Characteristics of Lithium-Ion Cell/Module from Various Manufacturers

Manufacturer	Cell/Module	Voltage (V)	Capacity (A·h)	Weight (kg)	Specific Energy (Wh/kg)	Power Density (W/kg)	Cycle Life	Application
CITIC Guoan MGL	M	24	13	2.3	92	151	500	Bike
	M	3.8	10	0.10	8	–	500	Phone
	M	3.8	5	0.5	76	125	500+	Miner lamp
	M	48	100	3.2	120	134	450	EV
Beijing Green Power	C	3.8	10	0.110	118	84	540	Bike
Tianjin Lishen Battery Joint-Stock Co., Ltd.	C	3.6	5.4	0.205	128	92	600	Notebook
	C	3.6	13	0.360	117	91	500	Bike
	C	3.6	8	0.245	121	94	500	Bike
Tianjin Lantian	C	3.6	2	0.044	–	–	–	Comm
	C	3.6	18	0.80	115	100	800	EV/bike
	C	3.6	100	2.6	106	138	460	EV/bike
	C	3.6	3	0.125	87	147	450	Power tools and HEV
	M	24	20	10.0	52	53	400	EV/motor bike
Xingheng	M	–	15	0.88	63	1,250	450	HEV/EV
	M	–	7.5	0.44	68	1,800	450	HEV
	M		10	0.37	100	200	500	EV
Suzhou Phylion Battery Co., Ltd.	C	3.7	10	0.36	102	86	500	E-bike
Shenzhen BAK Battery Co., Ltd.	C	3.7	1.8	0.054	123	100	–	Portable
	C	3.7	0.95	0.028	125	99	400+	Portable
ABT, Inc.	C	3.7	5	0.270	69	76	>700	Portable
GBP Battery Co.	C	3.7	60	1.80	123	82	>550	EV
Hyper Power Co.	C	3.7	1.15	0.037	115	78	>600	Portable
Shenzhen B&K Technology Co., Ltd.	C	3.8	0.100	0.007	100		400	Portable
Tianjin Blue Sky	C	3.7	60	1.8	122	80	>500	EV/bike
BYD Company, Ltd.	C	3.7	1.8	0.046	144	98	>400	E-bike
EMB Battery Co.	C	3.7	2.1	0.045	172	88	300	E-Bike

Most Chinese companies producing lithium-ion batteries do so for portable applications. Large companies have undertaken research and development with the help of joint ventures and/or partnerships with companies from Japan, Europe, and the United States. Section 3.4 provides summaries of joint ventures and partnerships. These companies, which include BYD Company Limited; EMB; GBP; Suzhou Phylion Battery Co., Ltd.; Xingheng; Tianjin Lantian; Tianjin Lishen Battery Joint-Stock Co., Ltd.; Beijing Green Power; and CITIC Guoan MGL, are developing lithium-ion batteries for e-bike, EV, and HEV applications — with particular focus on EVs and e-bikes.

E-bikes have been by far the most successful battery electric vehicle application in history, with an estimated cumulative production of ~30 million by 2007. At the heart of e-bike technology is the rechargeable battery. The core rechargeable battery technology used in e-bikes is VRLA, or "sealed lead-acid," and lithium-ion batteries.

Lithium-ion battery packs for e-bikes range from 24 to 37 V and have capacities of 5–60 A•h. The market for lithium-ion e-bikes in China is still small. In Japan and Europe, however, lithium-ion and NiMH are the dominant battery types, although annual e-bike sales (200,000/yr and 100,000/yr, respectively) are significantly lower than those in China.

In terms of traction battery technology, great achievements have been made in NiMH and lithium-ion/lithium-polymer battery technology. Research teams are:

- Beijing Powertronics Battery Co., Ltd. (lead-acid batteries)
- Tianjin Peace Bay Co., Ltd.
- Inner Mongolia Rare Earth Ovonic High-Power Ni/MH Battery Co., Ltd.
- Shenzhou Science
- Chunlan (NiMH battery technology)
- Zhong Hengrun
- Tianjin Lantian Double-cycle Tech Co., Ltd.
- Phylion Battery Co., Ltd.
- Beijing Green Power Technology Co., Ltd.
- MGL
- Thunder-sky (lithium battery technology)

3.1. Process to Identify Companies in China Developing Lithium-Ion Battery Technology

Initially, 32 web sites were identified as related to battery companies in China. These sites did not include company-specific sites. These sites were studied in detail to identify individual sites for battery companies, including those that mentioned such battery technologies as lead-acid, nickel-metal hydride, lithium-polymer, and lithium-ion. Approximately 260 company-specific sites were identified, and they included companies related to export, sales, research and development, and manufacturing battery technologies.

These 260 sites were studied in detail, and a hierarchy of battery technology criteria was established, as shown in Figure 1. For screening these companies initially, a set of "Selection Criteria" was set up to narrow down the list to companies involved in developing lithium-ion battery technology:

1. Sales
2. Research
3. Development
4. Manufacturing
5. Exporting
6. Size
7. Technology availability

On the basis of these criteria, 43 companies (see Appendix D, List A) were identified that are in business to pursue lithium-ion battery technology for various applications. These applications predominantly include camcorders, telephones, and a variety of electronic gadgets. Interest in technology development for transportation applications is strong and quite evident, particularly at large companies, universities, and institutions.

The number of appropriate companies was subsequently narrowed down to 27 (by using the criteria II shown in Figure 1) by eliminating sales companies (see Appendix D, List B). Five large institutions are also involved in the development of lithium-ion battery technologies.

In the next step, the exporting companies were eliminated. These companies do not develop any products; they simply buy products from other companies and export them. Also, extensive discussions were held with representatives from the China Automotive Technology and Research Center (CATARC) to narrow down of the final list of companies, institutions, and universities for the detailed analysis and evaluations. These organizations focus their work on lithium-ion technology and are shown in Appendix D, List C.

Section 3.2 provides details on lithium-ion battery technologies at several companies and institutions that were visited.

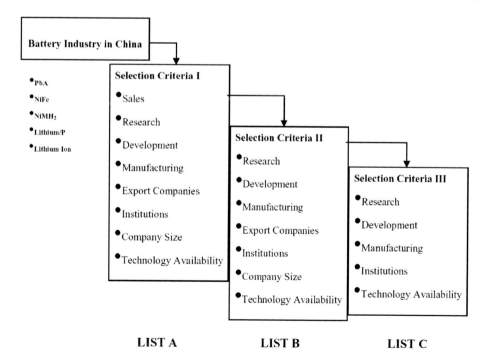

Figure 1. Process Used to Identify Companies in China Involved with Lithium-Ion Batteries.

3.2. Technology at Various Institutions in China

3.2.1. Tsinghua University (Department of Automotive Engineering Beijing, 100084)

The university's Automotive Department started its research activities for EVs in 1995, for hybrids in 1998, and for fuel cell hybrid vehicles in 1999. The university is working with the following battery companies on lithium-ion batteries for vehicular applications:

- Shenzhen B&K Technology Co., Ltd.
- Thunder Sky
- MGL New Energy Technology Co., Ltd.
- Oriental Polymer
- Huanyu Battery Co.

The department has an excellent facility to test and evaluate complete vehicular systems, as well as batteries at the module and pack level. Battery testing and evaluations are conducted by using the following:

- National Standards of the PRC
- USABC Battery Test Manual
- Partnership for a New Generation of Vehicles Battery Test Manual
- FreedomCAR Battery Test Manual
- Testing Standards of Japan
- Other testing standards, such as those developed by American Society of Mechanical Engineers (ASME)

Currently, it is evaluating hybrid and fuel cell hybrid buses with lithium-ion batteries. Batteries are 100 A·h, and a pack contains 30 cells. This evaluation is conducted for China's 863 Program. Oriental Polymer Company in Beijing is supplying an 1 8-kW fuel cell. Thunder Sky battery Company has provided a 30 A·h lithium-ion cell with 100 cells per pack in series.

A hybrid bus has been evaluated for 1,250 km without any degradation in battery performance. Similarly, a fuel cell hybrid bus has been evaluated for 2,340 km without any degradation in performance.

Other research includes:

- EV Structure Design
- Parameter Match and Optimization of EV Powertrain System
- Optimization of Energy Management Strategy
- EV Controller Design
- EV Communication Network
- EV Failure Diagnostics
- Other EV Subsystem Testing
- EV Assembly and Road Test

The department's focus is research on EVs, HEVs, PHEVs, and fuel cell electric vehicles (FCEVs), with an emphasis on battery applications. Involved in this research are 2 professors, 4 associate professors, 2 engineers, 5 part-time experts, 4 post-doctoral students, 4 doctoral candidates, and 13 master's degree students. The university has six patents and four applications pending in China on battery thermal management, EV controller design, and electronics for vehicles. The department is working with General Motors on fuel cell vehicles; however, the details of the involvement are unclear.

3.2.2. China Electrotechnical Society (46 Sanlihe Road Beijing 100823)

The China Electrotechnical Society has 123,000 members. The society is a clearinghouse for electrotechnical research. It conducts studies on battery technology markets for EVs and HEVs. Recently, the society completed a preliminary technology study on fuel cells and fuel cell hybrid vehicles. The study was conducted for its membership and the Chinese government. The society would not provide a copy of the report.

The Chinese government encourages the development of lithium-ion battery technology for vehicular applications. The Chinese government is providing incentives and grants for Chinese-owned and -operated companies — as much as 75% of the total cost, which covers most operating and capital expenses. For small companies, these incentives can be upwards of 85%.

China wants to export not only lithium-ion batteries for electronics applications, but it also wants to export batteries and advanced vehicles. Currently, China has 18 million battery-operated bicycles, of which 40,000 operate on lithium-ion batteries, and the rest on lead-acid batteries.

3.2.3. CITIC Guoan MGL MGL New Energy Technology Co., Ltd. (Beijing 102200)

MGL is located in Beijing Zhongguancun Science Park. It is engaged in the R&D and production of new composite metal oxide materials and high-energy-density lithium-ion secondary batteries. MGL is primarily supported financially by the CITIC Guoan Group, a wholly owned subsidiary of China CITIC Group. The CITIC Guoan Group is involved in such industries as information, new materials, comprehensive mineral resources exploration, tourism, and real estate. CITIC was founded in October 1979 by Rong Yiren, former Vice Chairman of the Peoples Republic of China. Through more than 20 years of development, CITIC has built a large-scale international enterprise group with total assets of 700 billion Ren Min Bi (RMB) ($933.34 million).

MGL ranks as China's largest manufacturer of the lithium-ion battery cathode material, $LiCoO_2$, and it will be the first to market the new cathode materials, $LiMn_2O_4$ and $LiCoO_{0.2}Ni_{0.8}O_2$. MGL states that it emphasizes quality control, and has passed the certification of both New and Hi-Tech Enterprise Standards and ISO9001:2000. With its particular synthesis method, MGL appears to efficiently produce cathode materials of superior performance and reliability in an environmentally friendly way. Since incorporation, MGL holds the monopoly in China's lithium-ion battery cathode materials market and now appears to be at the forefront of the industry. Besides cathode materials, MGL also produces lithium-ion secondary batteries of high energy density and high capacity for power and energy storage — the capacity ranges from several ampere-hours to several hundred ampere-hours. As China's first power-battery manufacturer, MGL appears to lead the industry in the marketing of high-capacity lithium-ion secondary batteries, used in the Beijing Municipality's trial electric bus fleet.

The MGL R&D center is located in Beijing. MGL has 850 employees, including 103 Ph.D. engineers and scientists. The R&D center encompasses 5,000 m^2 and includes advanced battery analytical, manufacturing, testing, and assessing equipment.

Major products of MGL's materials division include $LiCoO_2$ (2,300 tons/year), $LiMn_2O_4$ (1,450 tons/yr), and $LiCo_{0.2}Ni_{0.8}O_2$ (340 tons/yr). These oxide materials are indispensable to high-voltage (4-V) and high-energy-density lithium-ion secondary batteries. Over the past 10 years, lithium-ion secondary batteries have taken the place of NiMH and NiCd secondary batteries, which are used widely in mobile phones, laptops, and other electronic applications.

Instead of the commonly used solid-state synthesis method, MGL has adopted a unique method to synthesize materials, which is an efficient and simple process featuring zero emissions and low energy consumption. Feedback from lithium-ion battery manufacturers in China and abroad indicates that MGL's battery cathode materials have excellent and steady electrochemical performance. However, despite the rapid development of lithium-ion batteries over the last 10 years, limited cobalt resources and the poor thermal stability of $LiCoO_2$ limit the application of lithium-ion batteries.

For the past 10 years, with the support of the State and local governments, MGL has been focusing on the research and development of new lithium-ion battery cathode materials, notably spinel $LiMn_2O_4$ and layered $LiCo_{0.2}Ni_{0.8}O_2$. Experiments indicate that the superior thermal stability and steady charge-discharge performance of $LiMn_2O_4$ and $LiCo_{0.2}Ni_{0.8}O_2$ are quite compatible with various types of lithium-ion batteries. Recently, MGL's $LiMn_2O_4$-based and $LiCo_{0.2}Ni_{0.8}O_2$-based lithium-ion batteries have been used in energy-saving and environmentally friendly industries. China is poor in cobalt resources, but rich in manganese and nickel resources. MGL is developing China's lithium-ion battery materials industry by applying its specific synthesis method.

In comparison with NiMH and NiCd secondary batteries, lithium-ion secondary batteries have higher cell voltage, are smaller and lightweight, offer more flexibility under different temperatures, and do not have memory effects. Reportedly, these batteries do not emit pollution. At present, lithium-ion secondary batteries have an energy density of two to three times that of lead-acid batteries and around twice that of NiMH and NiCd batteries. Small lithium-ion batteries have been widely used in small high-end electronic devices (such as mobile phones and laptops). The physical-chemical properties of cathode materials, separators, and electrolytes play an important role in the reliability of lithium-ion batteries, and carbon — as the current anode material — has realized only one-tenth of its theoretical capacity. The new organic, inorganic, and metallic materials that MGL is developing will improve the physical-chemical performance of lithium-ion batteries and expand their applications, and solid and inorganic electrolytes will drastically improve the reliability and safety of lithium-ion batteries.

Recently, MGL's independently developed high-capacity lithium-ion batteries have been used successfully in Beijing's trial fleet of electric buses. Research indicates that the new lithium-ion batteries could accelerate the industrialization of EVs, and show potential in such applications as mobile communications, nighttime power storage, wind- and solar-power storage, backup emergency power, backup power for vehicles, and portable power.

3.2.3.1. Attributes of MGL's Lithium-Ion Battery

- Safe, excellent cycle performance and economics: the cathode material of MGL's power battery is spinel $LiMn_2O_4$, produced by MGL's unique synthesis method, which is yielding a far more economical and thermally stable product than $LiCoO_2$.
- Flexible configuration with aluminum packaging and stainless-steel casing.
- Stable output under large charge/discharge currents and different temperatures (-25~50°C).
- High voltage (3.8 V) and high energy density (>120 Wh/kg).
- No memory effects and non-polluting.

3.2.3.2 100-A•h $LiMn_2O_4$-Based Battery for EVs

- Constant current charge capacity (96%)
- Lower impedance (<0.85 mΩ)
- Flexible configuration
- Prominent safety performance
- High specific energy (>120 Wh/kg)
- Easily connected in series or parallel

3.2.3.3 Lithium-Ion Battery for Miner's Lamp

- High safety performance
- Lightweight, high specific energy
- Flexible capacity (4, 5, and 6 A•h)
- Long cycle life

In addition, MGL's LP 188270 series battery has passed the safety test by North Automotive Quality Supervision Test and Assessment Research Institute 863 (this is a National Advanced Technologies Development Program in China) Power Battery Testing Base under the standard GBIZ 18333.1-200 and GB/T 18827-2000. The testing results show that MGL batteries have superior safety performance, making the lithium-ion battery the ideal power source for a miner's lamp. Some manufacturers of miner's lamps in China have used MGL's batteries.

3.2.3.4. Battery-Management System for HEVs

- Measurement of current, voltage (cell and total), and temperature
- Calculation and display of SOC and state of health
- Cell balancing
- First protection (battery-management system): over-voltage (cell, pack), current, temperature
- Communication: CAN (ECU), RS-422 (internal), RS-232 (monitoring PC)
- Pre-charge control

- Second protection: over-voltage (cell), short current protection by heat sink (ratio of the porous sink height to the jet nozzle width).

3.2.3.5. MGL Battery-Powered Electric Vehicles

- MGL participated in the Challenge Bibendum 2004 held in Shanghai, China, and the Challenge Bibendum 2006 in Paris, France. At both events, MGL featured its independently converted pure EV powered by an MGL lithium- ion power battery. MGL received awards on both occasions for parameters such as acceleration, noise, fuel efficiency, radiation, and emissions.
- MGL's vehicles have passed the comprehensive EV performance tests performed by the Road Transportation Testing Center of the Road Science Research Institute of China's Ministry of Transportation.
- Beijing's Olympics electric bus, powered by an MGL 100-A•h power battery, has covered an cumulative range of 20,000 km during the past two years, and the tests have demonstrated high electrochemical and safety performance of its battery.

3.2.3.6. Lithium-Ion Power Battery Packs for Electric Two-Wheelers

- 24 V with 13-A•h aluminum packaged, weight 2.3 kg
- 24 V with 8-A•h stainless-steel packaged, weight 2.45 kg

3.2.3.7. Other Information on MGL

- MGL holds several patents (perhaps between 15 and 25) on cathode materials, anode materials, a battery management system, electronics, and other areas of battery technology in the United States, Japan, China, and Korea.
- The cost of the lithium-ion technology at the cell level is 5–9 RMB/Wh ($0.67–1.20/Wh).
- MGL is working with a U.S. company on a battery for electric and hybrid vehicle applications. However, MGL did not provide the name.
- MGL is receiving grants from local, district, and national government — the total amount is around 10 million RMB ($1.34 million) per year.
- MGL is willing to provide cells, modules, and packs for benchmarking in the United States.

3.2.4. Beijing Green Power Technology Co., Ltd. (Beijing 100086)

Beijing Green Power Technology Co., Ltd., has set up research laboratories, as well as applied laboratories for technology scale-up. In the research laboratory, the company conducts research on the materials for cathodes, anodes, and separators. In applied materials, the laboratory conducts research on cells and module development and performs testing and evaluation.

After three years, Green Power has formed its own technical research direction on the aspects of lithium-ion power cell technology and cathode material with technological success. Green Power has invested 20 million RMB ($2.67 million) to set up a product line of power cell capacity of about 20,000 A·h per day. Green Power has invested 10 million RMB ($1.34 million) to set up a product line of spherical lithium cobalt oxide cells with a capacity of 300 tons per year. Additionally, 15 million RMB ($2.00 million) was invested in setting up a product line of three-element cathode material with 300 tons per year, which was put into production in July 2006. The production capacity is 10,000 A·h per day.

3.2.4.1 Lithium-Ion Battery Technology

Electric Bike: Green Power uses its specific technology to fabricate cell structures, along with a new cathode material (LiNi1/3Co1/3Mn1/3O$_2$) and a new electrolyte (LiBOB), for its high- power cells and modules. The company has developed a 10-A·h cell for e-bike application. The cost of this cell is estimated at 135 RMB ($18.00) (this is only the cell cost). The e-bike battery pack has 10 cells (10 A·h each), and so far the company has sold 540 e-bikes with this battery and accumulated 30,000 km of on-the-road performance without battery degradation.

Electric Bus: A 100-A·h cell has been designed for the electric bus application. In the electric bus, the battery pack has 500 cells with 100-A·h capacity for each cell. This bus is being evaluated on the road and has accumulated 20,000 km; battery capacity is at 98% of the original capacity. A battery pack has 400 V and provides 50 A·h of capacity. Total energy is greater than 200 kWh. The pack is designed to operate at -20°C to +55°C. Green Power is working on a cooperative program with Beijing Science and Engineering University, which is evaluating electric buses for the 2008 Olympics. Green Power has built five buses so far. A thermal management system used in this bus maintains a temperature with a differential between 1 and 3°C. The cost of a 100-A·h cell is estimated at 1,700 RMB ($226.67). There are additional costs for making a battery pack. Green Power will not elaborate on how much cost is added to the cell cost.

Hybrid Electric Bus: For the HEV bus application, Green Power is using 100 cells, each with a capacity of 100 A·h. Each cell weighs 3.0 kg. Pack voltage is 360 V, and the power is 54 kW (with energy around 34 kWh). A hybrid bus that is similar to a U.S. 40-ft version has accumulated 10,000 km with very little degradation in performance. Green Power is cooperating with Tsinghua University on this program. According to Tsinghua University, Green Power's battery design and performance are comparable with those of other lithium-ion batteries evaluated by the university. Green Power has built six HEV buses so far.

3.2.4.2. Other Information on Green Power

- Green Power has 10 patents in China on cathode and anode materials, battery management systems, electronics, and other areas of battery technology. It is currently marketing in China; however, Green Power wants to expand its market abroad in the near future.

- Green Power is obtaining 10 million RMB ($1.34 million) per year in grants from the Chinese 863 Program, which is an advanced technology program.*
- Green Power has a strong desire to work with a company in the United States.
- Green Power is very interested in providing cells, modules, and packs for benchmarking in the United States.

3.2.5. Tianjin Institute of Power Sources (Tianjin 300381)

Dr. Wang Ji Qiang, Chief Engineer and Professor at the Tianjin Institute of Power Sources, is well known in China and is involved in the 863 Program as it relates to advanced battery development for the Beijing Green Olympic Games. He serves on several technologically influential committees with the Chinese Government, works closely with lithium-ion battery development companies, and knows the technology across the country. He also looks after the activities of the "Testing Center of Chemical and Physical Power Sources of Ministry of Information Industry."

Tianjin Institute of Power Sources is one of the two national laboratories involved in battery testing and evaluation activities and programs. This testing center of chemical and physical power sources of the Ministry of Information Industry of China was established in 1985. It is claimed to be the largest, most comprehensive, most authoritative independent quality-testing center for chemical and physical power sources.

The testing center of the Chemical and Physical Power Sources of China Ministry of Information Industry (QSTC) is attached to the Tianjin Institute of Power Sources. QSTC is located in Nankai District, Tianjin City, occupying 5,500 square meters. The testing center has 500 items of advanced testing equipment and instruments and more than 30 million RMB ($4.00 million) in fixed assets. At present, 76 staff members are in the center laboratory, and 48 of those are professors and engineers. The QSTC center laboratory undertakes programs of the 863 Program, carrying out the tasks of EV battery supervision and test technology. Many types of nickel metal hydride batteries, lithium-ion batteries, and super capacitors used in EVs have been tested and evaluated at QSTC.

The QSTC center laboratory has been authorized by the National Authorization Committee (#L0591) and China National Certification and Authorization Supervision and Management Committee as a qualified laboratory for calibration and measurement. This center is one of four laboratories in the world qualified to calibrate photovoltaic scales. It is also the government-appointed laboratory to test imported and exported battery-related products (as well as those used for passenger trains). The center laboratory is the consigned organization for the testing of storage batteries for Underwriters Laboratories (UL) certification and is the appointed laboratory of the China Quality Authentication Center. QSTC has acquired the hardware and software of China compulsory 3C product certification (voluntary) and is able to offer relative technical service and support.

---

* In March 1986, Deng Xiaoping launched China's National High-Tech Research and Development Program (also known as the "863 Program" — see www.863.org.com). The overall goals of the 863 Program are to:
  - Bridge China's gap against developed countries in high-tech frontiers by pooling resources in selected high-tech fields.
  - Drive science and technology advances in other relevant fields.
  - Produce high-caliber technological R&D talents.
  - Create opportunities for formation of high-tech industries and lay foundations for realizing national higher-level economic and social development by around 2000.

The Center's quality guarantee system is based on the requirements of ISO/IEC guideline 17025:2005, "Universal Competence Requirements for Testing and Calibration Laboratory." QSTC has also been approved as a National Defense Science and Technology Laboratory by the National Defense Science and Technology Industrial Laboratory Committee.

The Center appears to have excellent testing equipment and to follow international testing standards. It is able to test basic electro-chemical performance, environmental evaluation, and safety for chemical and physical power sources by using standards established by the IEC, UL, United Nations, Japan Institute of Standards, and the military. The power sources can be Zn-Mn, lithium-ion, silver-zinc, Zn-air, and lead-acid batteries series. NiCd, alkaline, nickel metal hydride, special kinds of batteries, silicon solar cells, thermoelectric generation series, thermoelectric cooling modules, and sensors can also be evaluated at this Center.

Since its foundation, QSTC has completed many tests and experiments (such as national testing, supervision of selective examination, performance evaluation, product grading evaluation, quality testing, comparison experiments, product judicial judging, technical judging, arbitration testing, export and import testing, and feasibility evaluation). At the same time, the center laboratory is conducting research on various power sources charge/discharge equipment, testing technology, and performance evaluation methods. The Center is actively providing technology support and consultation to its clients to accelerate the quality of chemical and physical power sources in China. QSFC has had extensive connections with counterparts at home and abroad. The Center keeps the testing results from clients confidential.

The Center has the following testing and evaluation equipment:

- Pulsed solar simuslator (German made)
- Automatic absorption spectrometer; made by Perkin-Elmer (United States)
- Digatron (charge/discharge instrument) (German made)
- Environmental testing equipment
- High-power charge/discharge instrument (German made)
- Arbin charge/discharge instrument (U.S. made)

Since 2001, a number of national projects have focused on the development of lithium-ion batteries and related materials. There are two major basic research programs:

1. High-energy-density lithium batteries
2. R&D on new materials for lithium-ion battery applications

The most important part of these two programs is a major technical innovation program on the development of advanced batteries for EVs and HEVs, as well as for fuel cell electric vehicle applications. There are also many programs supported by different state agencies and local governments. These programs are related to basic and applied research on new battery systems and new materials, as well as to pilot or mass production and trial applications of newly developed materials and advanced batteries.

3.2.5.1. Progress in R&D of Advanced Materials for Lithium-Ion Batteries

In 2001, a National Major Program on R&D of advanced rechargeable battery systems with a specific energy of 200–250 Wh/kg was set up and has since been conducted by a joint research group that includes Wuhan University, Tianjin Institute of Power Sources, Beijing Physical Institute of China Academy, and Harbin Engineering University. To achieve the above goal of specific energy, the requirements for a 3.5–4.5-V system were set up as follows: (1) advanced negative material specific capacity greater than 450 mA·h/g and more than 85% efficiency for first charge-discharge cycle and (2) advanced positive material specific capacity greater than 200 mA·h/g.

Possible candidate materials were selected, and some of them were thoroughly investigated. Evaluations encompassed new material synthesis, property evaluation, modifications, and performance of materials in the prototype cells. Among those explored further were Li-Ni-Co-O and Li-Ni-Co-Mn-O systems for positive materials and carbon. C-Si composite and lithium-alloy systems were evaluated as negative materials. Li/S or polymer-S systems with lower voltage (below 3 V) were also investigated. Typically, some nano-size and surface-treatment concepts were used for developing advanced battery materials. An optimum combination of Li (Ni, Co, Mn) -positive material with a specific capacity of 220 mA·h/g and C-Si composite negative material with a specific capacity of 550 mA·h/g and 91% efficiency for first cycle was demonstrated in both 18650 and soft packaged cells. These cells exhibited specific energy between 236 and 267 Wh/kg.

3.2.5.2. Progress in Developing Lithium-Ion Batteries for EVs, HEVs, and FCEVs

In 2001, Major National Programs on R&D of advanced rechargeable battery systems for EVs, HEVs, and FCEVs were initiated. These programs are part of the Chinese 863 Program (Electrical Project). R&D of high-energy-type (capacity range of 50–200 A·h) lithium-ion batteries for EV and high-power (capacity of 8–40 A·h) lithium-ion batteries for HEVs and FCEVs were developed. Liantian Company, Xinyuan Company, Leitian Company, and Beijing Institute of Non-ferrous Metals are involved in the development of these batteries. The National Testing Center reported a good deal of progress in evaluations related to specific energy, cycle life at deep depth of discharge, and simulation of vehicle running conditions. High-power, 15-A·h batteries are used in fuel cell/battery hybrid cars for demonstration. These batteries have been crash tested according to a national standard, and no hazard was observed.

3.2.5.3. Progress in Developing New Lithium-Ion Batteries for E-Bike Applications

D-size cells with 5-A·h capacity and 10–15-A·h prismatic cells with Al-doped $LiMn_2O_4$ and $Li(Ni_xCo_yMn_z)O_2$ positive are being produced for e-bike applications. Also, high-power 18650 and 26650 cells with $Li(Ni_xCo_yMn_z)O_2$ and $LiFePO_2$ positive materials have been successfully developed and can run on a 10-h rate (C) of discharge.

3.2.6. Tianjin Lishen Battery Joint-Stock Co., Ltd. (Tianjin 300384)

Tianjin Lishen Battery Joint-Stock Co., Ltd., established in 1998, is located in the Tianjin Huayuan Hi-Tech Industry Park. The company occupies an area of 85,000 m^2 and employs 5,000 people. It has a capitalization of 600 million RMB ($80.00 million), and its total investment has reached 1.5 billion RMB ($200.00 million). Tianjin Lishen Battery Joint-

Stock Co., Ltd., has imported advanced automatic production equipment from Japan, so the production of lithium cells is completely automatic.

By 2004, the company had an annual production capacity of 200 million cells, which consist of more than 100 specifications, ranging from cylindrical to prismatic lithium-ion batteries. Relying on its intellectual property rights, with the innovation-oriented institution and supportive policies, Tianjin Lishen Battery Joint-Stock Co., Ltd., has become one of the largest manufacturers of lithium-ion batteries and possesses the most advanced technology in China. The Chinese Version of *Forbes* magazine ranked Lishen as number eight in its "List of Highest Potential Enterprises" in China for 2006.

Since its inception, Tianjin Lishen Battery Joint-Stock Co., Ltd., has been focusing on management — putting quality as the top priority and pursuing battery innovations. The quality and performance of the company's cells appears to be world-class. Tianjin Lishen Battery Joint- Stock Co., Ltd., has obtained the certification of ISO9001-2000, CE (Conformité Européenne; French for "European Conformity"), UL, and ISO14001. These certifications have paved the way for the company to capture international and domestic market shares. Tianjin Lishen Battery Joint-Stock Co., Ltd., is supplying batteries to corporations such as Motorola ESG and Samsung. Recently, the company set up branches in North America, Europe, Korea, Taiwan, and Hong Kong in an effort to establish a powerful worldwide-marketing network. Its business is 60% export and 40% domestic.

Tianjin Lishen Battery Joint-Stock Co., Ltd., is said to have always valued technology research highly, and it has increased its investment in R&D. Recently, it has set up a Postdoctoral Workstation and a National Technology Center. In 2005, it has established a world- class Safety Test Center. These efforts ensure continuous development of quality battery products. Tianjin Lishen Battery Joint-Stock Co., Ltd., recently developed a new type of lithiumion battery, and its cathode material is LiFePO4 from Valence. Battery performance is good and safe. This battery is being evaluated by CATARC in battery electric vehicles. It also makes 18650 cells at this plant.

Tianjin Lishen Battery Joint-Stock Co., Ltd., makes both prismatic and cylindrical cells. The prismatic cells are used in mobile phones, portable PCs, digital cameras, MP3 players, and other high-technology devices. The cylindrical cells are used in portable PC, E-book, and satellite communication devices; digital video cameras and portable DVD players; and portable printers/portable scanners. High-power prismatic cells are designed for e-bikes, EVs, HEVs, and power tools.

Last year, the company sold 33,000 lithium-ion batteries for e-bikes and provided batteries for seven electric buses, four hybrid buses, and two fuel-cell-battery hybrid buses. These vehicles are now being evaluated at various universities and test centers.

Tianjin Lishen Battery Joint-Stock Co., Ltd., has 22 patents on all aspects of lithium-ion battery technology, as well as on manufacturing techniques. These patents are held in China, the United States, Japan, Europe, Korea, and Taiwan.

The Chinese government has awarded Tianjin Lishen Battery Joint-Stock Co., Ltd., with an initial 50 million RMB/year ($6.67 million/year) for three years and 80 million RMB/year ($10.76 million) for two years of subsequent work in manufacturing. Funding was possible because the company is located in Tianjin Huayuan Hi-Tech Industry Park, which is a declared tax-free zone.

The estimated cost of its cylindrical cell is 7 RMB/Wh ($0.93/Wh), and the estimated cost of the prismatic cell is 6 RMB/Wh ($0.80/Wh). The estimated cost of a high-power cell is 12 RMB/Wh ($1.60/Wh).

3.2.7. Tianjin Lantian Double-Cycle Tech. Co., Ltd. (Tianjin 300384)

In 2000, China Electronic Technology Group Corporations and Tianjin Metallurgical Group Co., Ltd., founded Tianjin Lantian Double-Cycle Tech. Co., Ltd., a high-tech enterprise. With capital of 93 million RMB ($1.24 million), it is specializes in research, development, and marketing of lithium-ion batteries and EVs. It is also engaged in research, development, and production of NiMH battery technology. The company has more than 500 employees, with an R&D staff of more than 150, including 30 postdoctoral fellows and 38 Ph.Ds.

The company has undertaken the EV project of the National 863 Program. A number of Tianjin's key projects include the development of lithium-ion batteries, an efficient electric machine, a controller, e-bikes, and EVs and associated parts. The company holds six patents, and six patents are pending. Tianjin Lantian Double-Cycle Tech. Co., Ltd., is certified by ISO 9001-2000, CE, and UL.

Tianjin Lantian Double-Cycle Tech. Co., Ltd., receives a 50 million RMB ($6.67 million) grant per year from the government. Its production facility is, at best, semi-automatic. Last year alone, the company sold 27,000 e-bikes with lithium-ion batteries at a cost of 3,000 RMB/bike ($400/bike). The company is working on an EV bus battery, and it has provided six 500-A•h batteries for testing and evaluation. These batteries are being developed within the National 863 Program.

Table 7. Tianjin Lantian Double-Cycle Tech. Co., Ltd., Cells for E-Bike Applications.

Cell Model #	Capacity (Ash)	Voltage (V)	Weight (kg)	Dimensions (mm x mm)	Operating Temperature (°C)	Cycle Life (80% DOD)
ICR 47/205	18	3.6	0.8	47 dia x 205	-20 ~ 55	800
ICR 50/380	55	3.6	1.8	50 dia x 380	-20 ~ 55	800
ICR 65/400	100	3.6	2.6	65 dia x 400	-20 ~ 55	800

Table 8. Tianjin Lantian Double-Cycle Tech. Co., Ltd., Cells for Power Tools and HEVs.

Cell Model #	Capacity (Ash)	Voltage (V)	Rate Discharge (C)	Weight (kg)	Dimensions (mm x mm)	Operating Temperature (°C)	Cycle Life (80% DOD)
ICR 45/1 50	8	3.6	15	0.45	45 dia x 150	-20 ~ 55	800
ICR 25/50	1.8	3.6	15	0.055	25 dia x 50	-20 ~ 60	800
ICR 32/65	3	3.6	10	0.125	32 dia x 65	-20 ~ 60	900

Table 9. Details on Other Lithium-Ion Cells for E-Bikes.

Cell Model #	Capacity (A·h)	Voltage (V)	Time Discharge	Weight (kg)	Volume (L)	Cycle Life
36V/9A·h	9	36	1 h 48 min (5A)	2.8	3.5	600
24V9A·h	9	25	1 h 48 min (5A)	2	2	600

Table 10. Tianjin Lantian Double-Cycle Tech. Co., Ltd., Cells for Electric Motorcycles and Other Electric Vehicles.

Cell Model #	Voltage (V)	Max Discharge Current (A)	Weight (kg)	Volume (L)	Cycle Life
24 V/20 A·h	24	20	10	5.5	>800
24 V/55 A·h	24	60	17	10.5	900
36 V/20 A·h	36	20	12	7.0	900
36 V/55 A·h	36	60	23	13.5	800
288 V/100 A·h	288	150	230	115	800

3.2.7.1. Cell Production and Applications

Tianjin Lantian Double-Cycle Tech. Co., Ltd., produces three different cells as the high-energy and high-power lithium battery for EVs and cells for e-bike applications. Table 7 provides details on these cells.

Table 8 lists details on high-power lithium-ion batteries for power tools and HEVs.

Table 9 shows details on other lithium-ion cells for additional applications for e-bikes.

Table 10 provides details on lithium-ion cells for electric motorcycles and other EVs.

3.2.7.2. Other Information on Tianjin Lantian Double-Cycle Tech. Co., Ltd.

- The company wants to export its products and would like to have a relationship with U.S. companies.
- The company has designed a battery for a hybrid vehicle with a capacity of 7.5 A·h, 96 cells (12 cells per module). It has fabricated 20–30 packs for laboratory and in-vehicle evaluations.
- The company would like to provide batteries for benchmarking in the United States, and it believes that doing so will enhance its potential for cooperation.

3.2.8. Suzhou Phylion Battery Co., Ltd. (Jiangsu 215011)

Phylion is a battery technology corporation set up by Legend Capital Co., Ltd.; the Institute of Physics of the Chinese Academy of Sciences; and the Chengdu Diao Group. Phylion has 82 million RMB ($10.93 million) and a staff of more than 400. Phylion specializes in manufacturing and selling lithium-ion cells with high capacity and current. Phylion's technology is primarily used in defense, electric bicycles, lighting, portable electronics, medical equipment, and battery-operated tools.

As one of China's largest manufacturers of lithium-ion battery cells, Phylion's average volume is currently 36 million units per year. Phylion's high-capacity (10 A·h) and high-power (7.5 A·h) lithium-ion cells have successfully passed U.S. testing by Underwriters

Laboratories, which paved the way for exporting the technology to the industrial world. Phylion benefited from technical assistance from the Institute of Biophysics of the Chinese Academy of Sciences (for about 15 years), funding support by The National Program #863 and #973, and further assistance from the Development and Reformation Committee (for the last 12 years). This combined funding enabled Phylion to accumulate experience as a company and establish a solid technical foundation to develop lithium-ion battery technology. Phylion has a 10 million RMB ($1.34 million) grant from the federal government to conduct basic and applied R&D on lithium-ion battery technology for e-bike, EV, and HEV applications. Furthermore, internally developed patents have been applied in the development of lithium-ion technology and methods of production. The company holds 17 patents on various aspects of lithium-ion battery technology. Phylion uses $LiMn_2O_4$ as the cathode in the lithium-ion battery.

Table 11. Summary of Tests Involving Charging Electric Cores with Constant Current and Setting Upper Limit for Fixed Voltage.

Standard	Premise	Environmental Temperature (°C)	Charging Current	Experimental Process	Time Requirements	Result Requirements
War Industry	After charging, according to the standards	20 ± 5	0.2C 5A	Until the protective circuit functions	No	No explosion and burning
QB/T25022000 Standards of Light Industry	The cell in the state of complete discharging	20 ± 5	0.2C 5A	Make the protective circuit function	12.5 h	No explosion and burning
04 Department of Science and Technology [863 Program], Storage Battery of Electro-motion Bicycle	After charging, according to the standards; discharge for an hour	20 ± 5	1C 1(A)	The voltage reaches 5.0 V	Or charge for 90 min	No explosion and burning
National Standards GB/T 18287-2000	After charging, according to the standards	20 ± 5	3C 5 A	The upper limit of voltage is 10 V	After declining the peak value 10 min, end the experiment.	No explosion and burning
Underwriters Laboratories Standards	After charging, according to the standards	20 ± 5	Carry through the relative current and time, according to: $I_c = \frac{2.5C}{3(I_c)}$	The testing time cannot be shorter than 48 h	No explosion and burning

Note: C is the standard capability. Ic is the testing current.

Phylion's production facility is semi-automatic. All material mixing, cell size cutting, and cell tabbing are done manually. The production equipment is imported from Japan, with 90% financial assistance from state and federal governments. In 2006, Phylion sold 52,000 packs for various applications and sold 32,000 e-bikes. The company builds 50 packs of different types for EV and HEV applications. Phylion has cooperative programs with Shanghai Automobile Group, North Automobile Group in Beijing, and Tongji University in Shanghai.

At present, Shanghai Automobile Group is evaluating Phylion's lithium-ion battery in a fleet of passenger vehicles, which includes EVs, HEVs, and fuel-cell-battery hybrid vehicles. The North Automobile Group is evaluating batteries in electric buses, and Tongji University is conducting an evaluation of batteries in fuel-cell-battery hybrid vehicles.

Phylion has conducted several experiments to determine the safety of the lithium-ion battery and has provided some experimental data, which are summarized in the following sections.

3.2.8.1. Overcharging Experiment

The overcharging experiment involved charging the electric cores with the constant current and setting up the upper limit for the fixed voltage. The interior electric core will raise the dendrite growth in the cathode. The tests are summarized in Table 11.

Table 12. Summary of Tests Involving Nail Penetration.

Standard	Premise	Environmental Temperature	Steel Needle	Experimental Process	Time Requirements	Result Requirements
War Industry	After charging, according to the standards	20 ± 5	Φ 3mm	Strongly penetrate along the radial	No requirements	No explosion and burning
Standards of Light Industry iQB/T2502-2000	After charging, according to the standards	20 ± 5	2.5 ~ 5 mm	Penetrate in the direction that the center and electrode surfaces are vertical	Place more than 6 hours	No explosion and burning
2004 Department of Science and Technology [863 Program], Storage Battery of Electromotion Bicycle.	After charging, according to the standards	20 ± 5	φ 3 ~ 8 mm	Promptly penetrate in the direction to which the pole plate is vertical	The steel needle sticks in	No explosion and burning
Underwriters Laboratories Standards	After charging, according to the standards	20 ± 5	Bounce to the cells at the minimum acceleration of 75 g and the maximum acceleration of 125 ~ 175g within 3 ms on the face and side face of the cell			No explosion and burning, ejection 5 g

3.2.8.2. Experiment with Nail Penetration

Table 12 summarizes the tests involving experiments with nail penetration.

3.2.8.3. Thermal Impact

Table 13 summarizes the tests involving experiments with thermal impact.

Table 13. Summary of Tests Involving Thermal Impact.

	Premise	Velocity of Ascending Temperature	Upper Limit Temperature	Time Requirements	Result Requirements
War Industry	After charging, according to the standards	The battery oscillates circles for four cycles between -40 ± 2°C and 70 ± 2°C. Maintain the end temperature for 2 hours in each temperature environment. The mobile time of alternating the temperature cannot be longer than 1 minute; maintain temperature for 2 hours below 25°C.			No deformation, crack, and leakage after the test; normally charge and discharge
Standards of Light Industry B/T2502-2000	After charging, according to the standards	5 ± 1	110°C	60 min	No explosion and burning
2004 Department of Science and Technology [863 Program], Storage Battery of Electro-motion Bicycle.	After charging, according to the standards	5 ± 2	70 ± 2°C	20 min	No leakage, deformation, explosion, or burning
National Standards GB/T 18287-2000	After charging, according to the standards	5 ± 2	95 ± 2°C	30 min	No explosion and burning
Underwriters Laboratories Standards	After charging, according to the standards	5 ± 2	95 ± 2°C	10 min	No explosion and burning

3.2.8.4. Other Information on Phylion

- Phylion would like to export its products and have relationships with U.S. companies.
- Phylion would like to provide batteries for benchmarking in the United States to enhance its potential for cooperation.

3.2.9. Tongji University School of Automobile Engineering (Shanghai 201804)

In 2002, the College of Automotive Engineering was formally established in the Shanghai International Auto City in Jiading District. It was founded by merging the Automotive Engineering Department, the New Energy Center of Automotive Engineering, and the College of Automobile Marketing and Management, in accordance with the requirements of the Shanghai Automotive Industry.

The College has a staff of 64, of which 19 are full professors, 16 are associate professors, and 13 are lecturers. The College has 730 full-time undergraduate students, 124 master's degree students, and 27 doctoral students. There are 11 postdoctoral researchers in the postdoctoral mobile research center. In addition, the College has set up an internship program at the master'sdegree level with several automobile companies. The College has extensive collaborative programs with several universities in Germany and the United States.

The Advanced Technologies Laboratories have the following facilities:

- Vibration laboratory
- Acoustic laboratory
- Engine dynamometer laboratory
- Fuel cell testing and evaluation laboratory

- Advanced battery testing and diagnostics laboratory
- Vehicle testing track
- Vehicle diagnostics laboratory
- Large-scale software and analysis laboratory
- Testing and evaluation of vehicles laboratory
- Aero-dynamic laboratory

At the School of Automotive Engineering, Tongji University has world-class facilities to integrate advanced batteries and fuel cells in vehicles and conduct basic and applied research for the automotive industry. These testing capabilities are comprehensive and cover research, testing, and evaluation. The equipment and capabilities are impressive.

The School of Automotive Engineering is involved in extensive research with lithium battery development companies, fuel cell development companies, and domestic and foreign automobile companies.

3.2.9.1. Automobile Companies

- Shanghai FCV Powertrain Co., Ltd.
- Volkswagen
- China Automobile Association
- China Automobile Dealers Association
- German engineering company, IAV GmbH (Ingeniergesellschaft Auto und Verkehr)
- EDS PLM Solutions
- Toyota Motor Co.
- Nissan
- General Motors

3.2.9.2. Battery Development Companies

- Huanyu Group
- CITIC Guoan Mengguli Corporation (MGL)
- Beijing Green Power Technology Co., Ltd.
- Beijing Oriental Polymer New Energy Co., Ltd.
- Leitian Green Electric Power Supply (Shenzhen) Co., Ltd.
- Thunder-Sky Battery Co.
- Suzhou Polylion Battery Co.

3.2.9.3. Fuel Cell Development Companies

- Shanghai FCV Powertrain Co., Ltd.
- Wan Xiaing FC Company
- Hang Zhou Energy Co.

The School of Automobile Engineering has seven fuel cell battery hybrid vehicles from three different developers to test and evaluate on the dynamometer and on the track. Each of these vehicles has a 15-kW polymer electrolyte membrane fuel cell stack and 10-kWh lithium-ion batteries. Faculty would like to have these vehicles ready for the 2008 Olympics. Argonne staff had an opportunity to ride in and drive two of these vehicles. The vehicles rode smoothly, and acceleration was as good as that of U.S.-manufactured vehicles. One of the vehicles had accumulated 20,140 km, and the other had 12,300 km. The lithium-ion batteries in both of these vehicles performed well and have a capacity of around 95% that of the original. These vehicles are small compared with U.S. vehicles — they weigh approximately 950 kg and carry four passengers.

Tongji University is very well connected with federal and state governments. The President, Dr. Gang Wan, is very well connected with the federal Department of Science and Technology. Recently, Dr. Wan became the Minister of MOST. He is also well associated with the state government and local automotive industry. Tongji receives a 15 million RMB ($2.00 million) grant from the federal government and 10 million RMB ($1.34 million) from the state government. It also receives several million RMB from automotive companies (the university did not want to disclose the names of those companies).

3.2.9.4. Other Activities at the School of Automobile Engineering

- A car, called *Chao Yue I*, which means "surpass," was built by Shanghai Fuel Cell Vehicle Power Train Co., under the direction from Tongji University, Shanghai Automotive Industry Corp., and other city departments. A total of $4.57 million RMB, from a central government grant, was used to develop the new hydrogen-based fuel cell vehicle engine. An additional $9 million RMB was offered to the company to continue further development of hydrogen fuel cell vehicles through 2007. Fuel-cell-powered cars are viewed as the most suitable vehicle for family use in China because they do not produce any harmful emissions — a great advantage as the country turns to vehicular mobility.
- The Beijing Oriental Company Group and Tongji University are building a new hydrogen refueling station in Shanghai. Tongji University is responsible for the overall management, while Shell, which is funding part of the project, is working with the school on the design, construction, maintenance, and operations of the station. The Beijing Oriental Company is providing the engineering and procurement services needed to deliver the packaged hydrogen compression, storage, dispensing system, and trucked-in compressed hydrogen for the station. This station is due for completion by the end of the year and will refuel 3 buses and 20 cars. The project is part of the Chinese Ministry of Science and Technology's national program for the commercialization of fuel cell vehicles in the country.

Tongji University is seriously interested in benchmarking advanced battery technology in the United States. University staff members have agreed to discuss this idea further with their battery development partners.

3.2.10. General Research Institute for Nonferrous Metals, Institute of Energy Materials and Technology Lab of Li-ion Battery (Beijing 100088)

The General Research Institute for Nonferrous Metals (GRINM), established in 1952, is the largest R&D institution in the field of the nonferrous metals industry in China. Dr. Hailing Tu, a well-known expert on semiconductor materials in China, is the president. Since its establishment, GRINM has carried out more than 5,000 projects. After 50 years of development, research areas have expanded to include microelectronic and photoelectronic materials, rare and precious metals materials, rare earth materials, energy technology and materials, special alloy powder and powder metallurgy materials, superconductor materials, nanotechnology and materials, infrared optical materials, nonferrous metals processing technology, advanced mineral processing and metallurgy, nonferrous metal composites, materials analysis, and testing. GRINM is a comprehensive research institute covering a wide range of research areas. One R&D area in which GRINM is involved is the development of lithium-ion battery technology for e-bikes, EVs, and HEVs, as well as for other portable applications, such as cell phones, camcorders, and various other electronic devices.

GRINM has also established a broad technical exchange, cooperation, and trading partnership with counterparts in more than 30 countries and regions. The institution is always ready to cooperate with partners around the world for mutual growth and development.

GRINM is conducting basic research on the materials needs and requirements for high-energy and high-power lithium-ion battery technology. It has focused on nanotechnology and $LiMn_2O_4$ materials for the cathode, graphite for the anode, and PC+DC+DMC+1m Li P F6 liquid electrolyte and PP/PE/PP separator for the development of a lithium-ion battery cell. GRINM developed all materials in house except for the separator, which it imports from Japan and the United States.

Table 14. Properties and Data from GRINM Lithium-Ion Batteries.

Parameter	HP-40	HP-30	HP-25	HP-17	HP-10	HP-8
Nominal Voltage	3.6	3.8	3.8	3.8	3.8	3.8
Average Capacity (A•h)	40	30	25	17	10	8
Diameter x Height (mm)	Dia 56 x 258	Dia 56 x 228	Dia 47 x 341	Dia 47 x 288	Dia 47 x 180	Dia 47 x 165
Typical Weight (kg)	1.55	1.48	1.17	0.98	0.60	0.45
Volume (dm^3)	0.64	0.60	0.59	0.50	0.31	0.28
Specific Energy (Wh/kg)	110	80	85	70	85	65
Specific Power (W/kg) (10 s/50% DOD)	898	1,048	1,082	1,320	1,500	1,450
Imax Max Discharge Current (A)	400	300	250	270	200	200
Voltage Limits Charge (V)	4.2	4.2	4.2	4.2	4.2	4.2
Discharge (V)	2.8	3.2	3.2	3.2	3.2	3.2

3.2.10.1. GRINM's Lithium-Ion Cells

GRINM has designed several lithium-ion cells; some properties and data are given in Table 14.

In some test results, GRINM showed that ampere-hour capacity dropped to 80% at the end of 800 cycles and to 60% at the end of 1,200 cycles. GRINM has fabricated 1,200 cells in each of the above categories.

At present, GRINM is working with Weifang Jade Bird Huanguang Battery Co., Ltd. This battery company will manufacture cells and batteries for EV and HEV applications, and these batteries will be integrated in the First Automotive Factory of China's electric and hybrid vehicles. GRINM has 10 e-bikes, two EVs, and two hybrid vehicles that are being evaluated.

The institute receives 5 million RMB per year ($0.67 million/yr) in grants from the federal government. It has 2,000 employees, and 38 are working on lithium-ion batteries. The Institute has seven patents, primarily on materials and cell design.

3.2.10.2. Other Information on GRINM

GRINM has an excellent record in terms of cooperating with other countries. The institute has shown very strong interest in a cooperative effort to benchmark its technology in the United States. Also, GRINM is interested in cooperating with other organizations to develop a lithium-ion battery for vehicular applications.

3.2.11. Beijing Institute of Technology BIT EV Center of Engineering and Technology (Beijing 100081)

The Chinese government has designated Beijing Institute of Technology (BIT) as the Center of Excellence for the EV Bus Development Program. BIT is the most prestigious institute in China. It has excellent laboratories with the most modern equipment for testing and evaluation of each component of an electric bus. Also, BIT can evaluate electric buses and passenger cars on a dynamometer and on controlled tracks.

A typical lithium-ion battery for a large-bus application consists of 108 cells in series and four banks in parallel to provide 400-A•h capacity with nominal 388 V. They charge cells to 4.2 V and discharge to less than 3 V.

3.2.11.1. Bus Manufacturing and Related Technology

BIT is working on bus manufacturing with the following companies:

- Beijing Beifang Huade Niopolan Bus Company, Ltd.
- Jinghua Bus Company, Ltd.
- BIT Clean Electric Vehicle Company, Ltd.

The technology that is being used for electric buses is described below.
Rare-earth charging flux permanent magnet direct current motor:

- Rare-earth permanent magnet and increasing magnetic winding combined excitation
- Rotor-adopting no-groove structure

- Increasing magnetic winding links to re-flowing current loop to auto-decrease magnetic field

Controller of rare-earth charging flux permanent magnet direct current motor:

- High-frequency PWM control
- Auto-decrease to realize the modulation
- Current of close loop controller
- Recovery of regenerative breaking energy

System parameters:

- System efficiency >92%, with 80% high-efficiency area for 84.4% of the time
- 75 kW/125 kW; maximum moment of system at 1,200 Nm
- Line-control two-speed gear box

Specifications for electric ultra-low-floor bus:

- Lithium-ion battery, 388.8 V @400 A·h
- Power driving system: three-phase, asynchronous alternating-current motor, 175 kW
- Wheelbase (mm): 5,800
- Wheel span (front/rear) (mm): 2,340/3,440
- Curb mass/full-load mass (kg): 12,930/16,000
- Max velocity (km/h): 91
- Driving range (40 km/h): 210 km
- 0–50 km/h acceleration time in s: 20.7
- 30 km/h braking distance in m: 8.2

BIT has 12 electric buses that are being evaluated. These buses will be put in use during the 2008 Beijing Olympics. Nearly all of the laboratory's equipment to evaluate electric buses has been imported from Germany, Japan, and the United States.

BIT is 100% supported by the Chinese federal and state governments. BTI has several patents on electric drive trains for buses, as well as for passenger vehicles.

3.2.11.2. Other Information

BTI is interested in working with U.S. companies on electric buses, as well as on component technologies.

3.3. Lithium-Ion Battery Manufacturing in China

The production and manufacturing of lithium-ion batteries has expanded enormously in the past two to three years. These batteries are widely used in consumer electronics devices, such as cellular phones, camcorders, and portable computers. This market is growing so fast that it is only second behind the development of information technology. Battery manufacturing ranges from manual operations to full automation. Labor cost also plays a significant role in manufacturing lithium-ion batteries. Currently, China has between 275,000 and 325,000 workers in the battery industry. Labor costs in China are very low compared with labor costs in the west and Japan. Approximately 435 batteries companies in China produce batteries. The total output in 2006 was 78.7 billion Yuan ($1.06 billion).

The Chinese government aggressively supports the development of new types of batteries, such as lithium batteries, fuel batteries (fuel cell), and solar batteries. China's 10th five-year plan and 863 Program listed the lithium battery and related key materials as an important research project in the new material field, which greatly promoted the research and development of the lithium battery. To realize the goal of holding a "Green Olympics" in Beijing and to welcome the Shanghai World Expos in 2010, Beijing and Shanghai launched a series of regulations, policies, and measures to promote applications of the battery for automobiles to reduce exhaust emissions and protect the environment. In October 2003, the National Development and Reform Commission, Ministry of Science and Technology, developed a solar energy R&D plan for implementation in the next seven years — the "Bright Project" of the National Development and Reform Commission will raise 10 billion Yuan ($1.35 billion) to promote applications for solar power generating technologies, and the installed capacity of China's solar power generating system is expected to reach 300 MW by 2009.

Because of the reduced use of NiMH batteries and nickel-cadmium batteries and the popularization of mobile electronic products such as mobile phones, digital computers, digital video, and personal data acquisition, the use of lithium-ion batteries surged, and the potential of these batteries in the rechargeable battery field is significant. Worldwide, the output of lithium- ion batteries surpassed that of nickel-cadmium batteries in 2003, and the trend continued in 2004.

The number of shipments of mobile phones and laptops continued to increase in recent years, although the growth of mobile phones slowed down in 2005. The growth, however, of laptops, DCs, and DVs enabled the rate of growth of the lithium-ion battery industry to remain high. The compound growth rate of world production of the lithium ion battery reached 23% during 2004–2006; the compound growth rate of sales revenue will reach 12.3%, according to MOST. During 2007–20 10, the lithium-ion battery industry will enter a stable growth period, during which the rate of growth in production is expected to be 9.85%, and the compound growth rate of sales revenue is expected to reach 5.85%, according to the Tianjin Institute of Power Sources.

The lithium polymer battery was mass-produced in 1999, and the growth rate of these batteries is always higher than that of the lithium-ion battery. The lithium polymer battery accounted for 7%, 8%, and 10%, respectively, of the market share for lithium-ion batteries in 2002, 2003, and 2004.

R&D trends of the lithium-ion battery industry include further increases in energy density; safer environmental performance; lower manufacturing cost; development of new electrode materials; development of lighter, thinner batteries; and development of improved manufacturing techniques.

3.4. Joint Ventures and Partnerships

Western and Japanese companies are setting up joint ventures and/or partnerships with Chinese battery companies because of the widespread incentives available to Chinese companies. Chinese companies receive more incentives if they are producing batteries for export. At present, Chinese companies are providing batteries for electronic and portable applications worldwide. Because the initial investment to produce lithium-ion batteries is very high, producing batteries in China is cost-effective.

The following paragraphs briefly describe the relationships of various western and Japanese companies with Chinese battery companies for production of lithium-ion batteries.

Shenzhen BAK Battery Co., Ltd., has signed manufacturing agreements with (1) Lenovo Group, Ltd., for portable and tabletop personal computers and cell phone markets and (2) A123 Systems for high-power lithium-ion battery technology based on patented nanotechnology. The Strategic Cooperation Agreement with Lenovo calls for both companies to jointly contribute and share resources to further product-development efforts. The company has been a supplier of lithium-ion battery cells for Lenovo's cell phones since last year (2005). This Agreement was signed to expand the current relationship and facilitate the development of new battery sizes for Lenovo in the portable and notebook PC markets. Shenzhen BAK Battery Co., Ltd., and A123 have collaborated since early 2005 to design, develop, and implement an advanced mass- production line to exclusively manufacture first products for A123. The A123 Systems nano- phosphate lithium-ion technology is based on patented technology developed at the Massachusetts Institute of Technology (MIT). The A123 high-power batteries will be used in a variety of applications, including power tools, medical devices, and hybrid electric vehicles. In 2005, Shenzhen BAK Battery Co., Ltd., raised approximately $60 million that enabled a significant expansion of manufacturing facilities, which increased production from 15 million to 22 million batteries per month. Shenzhen BAK Battery Co., Ltd., also established volume production capability for lithium-polymer battery cells and constructed a production line for a new high-power battery cell initially directed toward the cordless power tool market and other applications. Shenzhen BAK Battery Co., Ltd., has successfully launched its first automated cylindrical lithium-ion battery cell production line. The new production line employs advanced technologies and automation for the manufacture of consistently high-quality cylindrical cells for notebook computers. The monthly production capacity has reached two million units.

ARIA Investment Partners II, a fund managed by CLSA Private Equity Management, Ltd., has invested $10 million in China's Great Speed Enterprises, Ltd., a holding company of Scud (Fujian) Electronics Co., Ltd. Scud Electronics is one of the largest makers of lithium-ion batteries for mobile phones in China. Its retail distribution network has more than 200 wholesalers, which manage over 100,000 points of sales and cover all the provinces and large cities across China.

Advanced Battery Technologies, Inc. (ABAT), a subsidiary of Beijing Tonghe Jiye Trade, Ltd. (Tonghe Jiye), has received an order for 100,000 mine-use lamps using ABAT's 3.7-V, 9-A•h lithium-ion polymer batteries. This order has a total contract value of 24.8 million RMB (about $3 million). The Chinese government has given ABAT a safety certificate that allows the production of mine-use lamps using this battery. Mine-use lamps have an annual market of about $120 million in China. ABAT is resuming production in its newly built factory in Heilongjiang, China, with three production lines. The new factory has a daily lithium-ion polymer battery production capacity of 50,000 A•h per 8-h shift, which is 10 times the capacity of the preexisting plant. The old facility is being converted into an R&D laboratory. The company's products include rechargeable lithium-ion polymer batteries for electric automobiles, motorcycles, mine- use lamps, notebook computers, walkie-talkies, and other personal electronic devices. ABAT has filed a patent application for its nano-lithium-ion battery system. ABAT has been developing a new polymer lithium-ion battery by using lithium titanate spinel nanomaterials provided by Altair Nanotechnologies, Inc. ABAT has developed a method for incorporating the nanomaterials into its battery. The company has shipped its first group of new nano-lithium-ion batteries to its U.S. customer. The nano-lithium-ion batteries use lithium titanate spinel electrode nanomaterials provided by Altair Nanotechnologies, Inc.

Fengfan Storage Battery Co., Ltd., and the Hongwen Group jointly developed a lead-acid production facility in Guizhi, Tanshang City, of North China's Hebei Province, with an estimated investment of one billion yuan ($121 million). Fengfan contributed 51% of the investment, and Hongwen contributed the remaining 49%. The Fengfan-Hongwen storage battery project will mainly produce batteries for use in automobiles. This joint venture plans to develop lithiumion battery technology for portable and automobile applications.

Inco has officially opened its joint venture nickel foam plant — Inco Advanced Technology Materials (Dalian) Ltd. — in Dalian, China. Nickel foam is a specialty nickel product used in NiCd and NiMH rechargeable batteries, including batteries used as a power source for hybrid automobiles.

According to *China Customs* statistics, mainland China's battery production volume currently represents at least 25% of the global supply. Some industry sources say the percentage may even be as high as 50%. The China Industrial Association of Power Sources reported that 28 billion units of batteries were produced in mainland China in 2004. The 151 battery makers included in their report represent about 70% of mainland China's production capability. These have an individual output of between 5,000 units and 170 million units monthly, with about 96% of them operating at 70% of production capacity. Most of the battery makers are located in Guangdong, Shanghai, Zhejiang, Jiangsu, Tianjin, Shenyang, Fujian, and Harbin. Other findings in this report include:

1. Mainland China is the world's largest supplier of alkaline batteries, with output of primary cells reaching 22 billion units in 2004. Production was increased by 10% in 2005.
2. Over 800 million units of NiMH batteries were manufactured in mainland China in 2004. This is about 40% of the global supply in 2004. Exports of NiMH batteries totaled 675 million units, valued at $442 million in 2004. About 60 new companies included NiMH rechargeable batteries in their product lines in 2005.
3. About 30% of the world's requirement for sealed lead-acid batteries is supplied by mainland China. Shipments of sealed lead-acid batteries totaled about 107 million

units in 2004, which was an increase of 25% from 2003. In 2005 and 2006, electric vehicles, uninterrupted power sources, emergency lighting, security systems, and industrial applications drove the demand for sealed lead-acid batteries. Mainland China's production of NiCd batteries hit 1 billion units in 2004. Production is expected to grow by 5–10% in the next five years. Exports of NiCd reached 826 million units in 2004, valued at $411 million.

Ovonic Battery Company, Inc., has entered into a patent license agreement in connection with its proprietary NiMH battery technology with *Hunan Corun Hi-Tech Co., Ltd.*, of the PRC. Under the consumer battery license grant, Hunan Corun has a royalty-bearing, nonexclusive right to make, use, and sell NiMH batteries for consumer and nonpropulsion applications. Hunan Corun was founded in 2001. Its factory is located in the Chao Yang economic development zone in YiYang, Hunan, within the PRC. Hunan Corun manufactures a wide range of NiMH products, has approximately 3,000 employees, and produces a variety of battery chemistries. Ovonic also entered into a patent license agreement in connection with its proprietary NiMH battery technology with *Zhejiang Kan Battery Co., Ltd.*, of the PRC. Under the consumer battery license grant, Zhejiang Kan has a royalty-bearing, nonexclusive right to make, use, and sell NiMH batteries for consumer and nonpropulsion applications. Zhejiang Kan was founded in 1993 and is technically affiliated with *Zhejiang University*. Zhejiang Kan's factory is located in Suichang, Zhejiang, and its marketing office is in Hong Kong. Additionally, Ovonic has entered into a patent license agreement in connection with its proprietary NiMH battery technology with *L&K Battery Technology Co., Ltd.*, of the PRC. Under the consumer battery license grant, L&K Battery has a royalty-bearing, nonexclusive right to make, use, and sell NiMH batteries for consumer nonpropulsion applications.

Suntech Power has signed a letter of intent with Wanzhou District of Chongqing Municipality to construct a battery production base involving an investment of 200 million yuan ($24.9 million). Its local partner will be *Wanguang Power Source*, an old state-owned enterprise. The state-owned Assets Supervision and Administration Commission of Wanzhou District is negotiating with Suntech on acquisition details. Wanguang Power Source makes storage batteries used in automobiles and motorcycles, with annual output valued at 180 million yuan ($24.3 million). It is 51% held by Wanguang Industrial Group, 29.9% by Lifan, and 16.63% by Chongqing Xinmiao Technology Investment Company. Low labor cost is a factor that affects Suntech's domestic expansion scheme. Suntech, which is headquartered in China's eastern Wuxi City, is the sixth largest solar energy company in the world, with an annual battery output of 120 MW. Its products are mainly for export. Chongqing is a leading automobile- and motorcycle-manufacturing base in China, which also is an important reason for Suntech's plan to build a battery production base in the city.

ZAP is using ABAT's lithium-ion polymer batteries in its electric vehicles. Under the first phase of a new agreement, ABAT will retrofit a range of ZAP EVs with its lithium-ion polymer batteries and chargers. The initial testing shows that the ABAT batteries are increasing the run time of ZAP's vehicles by three times over that of lead-acid batteries. The threefold increase in energy density of the lithium polymer batteries could enable a similar threefold increase in transportation range for comparable-weight batteries, enabling ZAP's vehicles to achieve a significantly increased driving range between electric recharges.

Degussa AG, Düsseldorf, is starting additional electrode production for large-volume lithium-ion batteries at the Li-Tech GmbH (SK Group) site in Kamenz/Dresden. The first expansion stage saw anode and cathode production come onstream in the fourth quarter of 2006. Degussa has production capacities in China through the joint venture *Degussa-ENAX (Anqiu) Power Lion Co. Ltd.* The German site is targeting large-volume energy storage applications, such as batteries for hybrid vehicles. With its ceramic membrane, Separion®, the specialty chemicals company had positioned itself at an early stage in the promising market for lithium-ion batteries. According to Degussa, the global market for lithium-ion battery materials experienced double- digit growth in 2004 and currently amounts to more than $1.2 billion. Degussa expects the market volume to increase to around 4 billion by 2015.

Ultralife Batteries, Inc., has completed the acquisition of ABLE New Energy Co., Ltd., which is located in Shenzhen, China. Established in 2003, ABLE produces primarily Li-MnO$_2$ and Li-SOCl2 batteries for a wide range of applications worldwide, including utility meters, security systems, tire pressure sensors, medical devices, automotive electronics, and memory backup, among many others.

3.5. Government Policies

The Government policies provide tax and investment incentives for the following areas of advanced technologies research, development, and manufacturing:

Guidance on previously developed high-tech industrialized key fields: The National Development and Reform Commission, Ministry of Science and Technology, Ministry of Commerce of PRC, published Guidance on Prior-Developed High-Tech Industrialized Key Fields in April 2004. This content is associated with the battery industry and includes new types of batteries/power supplies such as lithium-ion cells and batteries, portable photovoltaic (PV) power supply systems, and a new type of solar film battery. The federal government provides additional incentives for the development of these technologies. This includes subsidies for hiring recent graduates and some tax break.

Materials with special functions: This category includes energy conversion and energy storage materials, including a hydrogen storage alloy and a hydrogen storage container, solar batteries, high-performance rechargeable lithium batteries, and a new type of capacitor.

Fuel battery (fuel cell): Fuel cells use hydrogen or rich-hydrogen gas as fuel and oxygen as oxidant; it converts chemical energy into electricity. This energy conversion is efficient and environmentally friendly because almost no nitrogen oxide or sulfur oxide is discharged. The key technologies included in the recent industrialization are battery materials; 1-kW–100-kW proton- exchange membrane fuel cells and electrical catalysts; electrodes; Nafion-polytetrafluoroethylene compound films; bipolar plates; proton-exchange membrane fuel cells; and direct-methanol, molten carbonate, and solid oxide fuel cells.

New energy and renewable energy: China possesses rich renewable energy resources. Properly developed and promoted, renewable clean energy (such as bio-energy, wind energy, solar energy, hydrogen energy, and geothermal energy) has great potential to reduce pollution and improve China's energy structure. Its recent industrialization plan also includes: biomass gasification; electricity-generating and air-feeding technology; biomass liquid fuel technology; design, manufacture, and production of a 750-kW wind-power generator set and related components; a megawatt wind-power generator set and related components; high-efficiency, low-cost solar PV cells; medium- and high-temperature solar-energy-generating equipment; a ground-source heat pump; and a heating, air-conditioning, and hot-water combined supply system.

3.5.1. Renewable Energy Law

On February 28, 2005, the Renewable Energy Law of the People's Republic of China was passed during the 14th session of the 10th National People's Congress Standing Committee. This law, effective as of January 1, 2006, will greatly promote the development and use of renewable energy, the increase in energy supply options, the optimization of energy structure, the energy security guarantees, and the environmental protection and sustainable development of China. The Renewable Energy Law defines the term "renewable energy" as non-fossil energy sources, which include wind, solar energy, hydropower, biomass, geothermal energy, and ocean energy.

3.5.2. Government Plan

The rechargeable lithium-ion battery is a new and important technology in the energy field and is strongly supported by the Chinese government. Since the initiation of China's 863 Program in 1987, the Ministry of Science and Technology organized the research and development of the key materials and technologies of the NiMH battery and lithium-ion battery that shaped a large industrial-scale role for Chinese companies worldwide.

During the period of the 10th five-year-plan, China could realize the batch production and industrialization of hybrid electric automobile and build power batteries and industrial bases for related materials. Recently, the National Development and Reform Commission of China decided to support the construction of a base from which to demonstrate the industrialization of lithium-ion batteries to further promote the development of lithium-ion battery.

Compared with other countries, China still lacks the ability to update technology rapidly and introduce innovations. The "973" Plan of the Ministry of Science and Technology has been approved to promote the development of green technologies and accelerate the development of the energy structures that are green and renewable. Establishing this green rechargeable battery technology program will emphasize innovative research in battery materials, new systems of green rechargeable batteries, and related technologies.

3.5.3. Relevant Standards

The national standards of the battery industry are summarized in Table 15.

A new battery standard, GB8897.2-2005, was issued in January 2005 and became effective on August 1, 2005, replacing GB/T71 12-1998. The original standards of the light industry, QB/T528-1966, QB/T1186-1991, and QB/T1732-1993, were abolished at the same time. In 2005, the product quality testing standard of the battery industry was changed to GB8897.2-2005.

The Standardization Administration of China will draw new standards for cell-phone use of lithium-ion battery to replace the existing ones. At the national standard coordination meeting held in Qingdao in 2004, Aucma New Energy Company was chosen to take charge of the development of new standards for lithium-ion batteries used in cell-phones. The draft work started in July 2004, which clearly prescribes the capacity, cycling life-span, and safety of batteries used in mobile phones.

Table 15. Guobiao (GB) Standards of the Battery Industry.

GB Number	Title/Name
GB/T 2297-1989	Terminology for solar photovoltaic energy system
GB/T 2900.11-1988	Terminology of (secondary) cell or battery
GB/T 5008.1-1991	Lead-acid starter batteries — Technical conditions
GB/T 5008.2-1991	Lead-acid starter batteries — Varieties and specifications
GB/T 5008.3-1991	Lead-acid starter batteries — Dimension and marking of terminals
GB/T 6495.1-1996	Photovoltaic devices — Part 1: Measurement of photovoltaic current-voltage characteristics
GB/T 6495.2-1996	Photovoltaic devices — Part 2: Requirements for reference solar cells
GB/T 6495.3-1996	Photovoltaic devices — Part 3: Measurement principles for terrestrial photovoltaic (PV) solar devices with reference spectral irradiance data
GB/T 6495.4-1996	Procedures for temperature and irradiance corrections to measured I-V characteristics of crystalline silicon photovoltaic devices, Battery Industry in China, 2005
GB/T 7169-1987	Designation of alkaline secondary cell
GB/T 7260-1987	Uninterruptible power systems
GB/T 7403.1-1996	Lead-acid traction batteries
GB/T 7403.2-1987	Lead-acid traction batteries — Product types and specifications
GB/T 7404-1987	Lead-acid batteries for diesel locomotives
GB/T 8897-1996	Primary batteries: General
GB/T 9368-1988	Nickel-cadmium alkaline secondary cells
GB/T 9369-1988	Nickel-cadmium alkaline secondary batteries
GB/T 10077-1988	Maximum outside size and capacity series for lithium batteries
GB/T 10978.2-1989	Lead-acid batteries for special type explosion-proof power unit in coal mines — Product types and specifications
GB/T 11009-1989	Measuring methods of spectral response for solar cells
GB/T 11010-1989	Spectrum standard solar cell
GB/T 11011-1989	General rules for measurements of electrical characteristics of amorphous silicon solar cells
GB/T 11013-1996	Alkaline secondary cells and batteries — Sealed nickel-cadmium cylindrical rechargeable single cells
GB/T 12632-1990	General specification of single silicon solar cells
GB/T 12637-1990	General specification for solar simulator
GB/T 12724-1991	General specification for silver-zinc rechargeable cells
GB/T 12725-1991	General specification for alkaline nickel-iron rechargeable batteries
GB/T 13281-1991	Lead-acid batteries used for passenger trains
GB/T 13337.1-1991	Stationary acid spray-proof lead-acid batteries — Technical conditions
GB/T 13337.2-1991	Stationary acid spray-proof lead-acid batteries — Capacity specifications and size
GB/T 13422-1992	Power semiconductor converters — Electrical test methods
GB/T 14008-1992	General specification for sea-use solar cell modules
GB/T 15100-1994	General specification for alkaline nickel-metal hydride cylindrical sealed rechargeable batteries
GB/T 15142-1994	General specification for nickel-cadmium alkaline rechargeable single cells
GB/T 6495.5-1997	Photovoltaic devices — Part 5: Determination of the equivalent cell temperature of photovoltaic (PV) devices by the open-circuit voltage method
GB/T 17571-1998	Alkaline secondary cells and sealed-nickel batteries — Cadmium rechargeable monobloc batteries in button cell design
GB 8897.2-2005	Primary battery — Part 2: Sizes and technical conditions

3.5.4. Regulatory Steps to Promote Cleaner Transport

In the new Five-Year Plan (2006–2010), the Chinese government outlines steps to boost efficiency and reduce pollution. A number of clear targets for increasing energy efficiency are set (e.g., to increase total energy efficiency by 20% and achieve an energy mix of at least 20% renewable energy by 2020). The Chinese government also introduces clear policies:

- As of April 1, 2006, buyers of new, big cars paid 20% more in sales tax, and buyers of smaller cars paid 1% less in sales tax.
- By the end of 2006, the parking fees in central Beijing doubled.
- During 2006, the price of gasoline for end consumers increased by about 20% (because older subsidies had been lifted).
- Starting December 2005, tighter vehicle standards were approved.

3.5.5. Example of Income Taxes and Incentives

Under applicable income tax laws and regulations, an enterprise ABC (not a real name of a company) located in Shenzhen, including the district in which operations are located, is subject to a 15% enterprise income tax. Further, according to PRC laws and regulations, foreign-invested manufacturing enterprises are entitled to, starting from their first profitable year, a two-year exemption from enterprise income tax, followed by a three-year 50% reduction in the enterprise income tax rate. The PRC subsidiaries are entitled to a two-year exemption from enterprise income tax and a reduced enterprise income tax rate of 7.5% for the three years following the first profitable year. As such, for the first two calendar years, ABC enterprise was exempted from any income tax. For the following two years, this enterprise is subject to the income tax rate of 7.5%. Some preferential tax treatment is also applicable to this enterprise, and the enterprise is fully exempt from any income tax during a tax holiday. (A tax holiday is a designated period — the month of June each year — during which companies do not pay income tax on equipment purchases or any other incurred business expenses.)

In addition, due to the additional capital invested in the ABC enterprise, it was granted a lower income tax rate of 1.7% for two years. Furthermore, to encourage foreign investors to introduce advanced technologies in China, the government of the PRC has offered additional tax incentives to enterprises that are classified as a foreign-invested enterprise with advanced technologies. If the enterprise qualifies for this designation, then it pays 1.7% in taxes for an additional three years. It can then renew this status and pay low income taxes. As a result, as long as ABC maintains this designation, it may apply to the tax authority to extend its current reduced tax rate of 1.7% for another three years.

3.6. Government Policies and Hev Standards

The HEV-pro policy has been clearly included in the new China Auto Industry Policy. The R&D and industrialization of HEVs have been listed as a major component of government funding for technology development in China. Also, the related HEV standards will be recommended and adopted by auto makers in China:

- Test methods for energy consumption of light-duty hybrid EVs
- Measurement methods for emissions from light-duty hybrid vehicles
- Program for approving the engineering evaluation of HEVs
- Programs for testing the comprehensive performance of HEVs
- Safety specifications for HEVs
- Method for testing the power performance of HEVs
- Technical specifications of the motor and controller for EVs
- Method for testing the motor and controller for EVs

4. SUMMARY AND RECOMMENDATIONS

The lithium-ion battery offers very high power on charge and discharge and further improvements — desirably in power at low temperature — may be possible. The main challenge for this technology — besides cost reduction — is to achieve acceptable operating life, particularly at 40°C. Battery companies, as well as R&D organizations worldwide, are making major efforts to mitigate the relatively rapid fading of the $LiMn_2O_4$ lithium-ion battery at elevated temperatures. However, the degree of improvement that will be achieved is difficult to anticipate.

The basic chemistry and design of lithium-ion HEV cells are quite similar to those of small consumer cells, which suggests that the basic manufacturing processes for HEV batteries should be well understood. The manufacture of lithium-ion cells is known to require a higher level of process control and precision than most other types of battery manufacturing, and, as a result, scrap rates tend to be higher. Most, if not all, producers of small lithium-ion batteries have experienced product recalls and/or production shutdowns as result of reliability issues and/or safety incidents. Extrapolating this experience to the much larger HEV cell with thinner electrodes, it seems likely that scaling up the production of HEV cells from the current early pilot level will be slow and costly. If lithium-ion HEV batteries are to become commercially viable, operating life and abuse tolerance issues will need to be resolved first, and then the unit cost of the technology will need to be reduced, at least to the levels projected for NiMH batteries.

Because the 2008 Olympics will be held in Beijing, the Chinese government designated lithium-ion battery technology as a strategic technology for the development and manufacturing of portable and vehicular applications. As a result of this designation, foreign companies were attracted to work with Chinese companies to form joint ventures and/or partnerships, since the potential size of Chinese markets for portable electronics and vehicular applications is significant.

The following recommendations are made:

1. A DOE Program Official(s), along with experts, should visit China to study its battery technology industry firsthand and to make arrangements for benchmarking Chinese battery technology in the United States. Chinese companies have expressed a strong interest in making battery technology available for benchmarking. The timing is right, and interest in working with the United States is very strong.

2. The DOE and the Tianjin Institute of Power Sources should work together to set up a battery workshop in China and invite U.S. industry to participate. This effort will help the U.S. industry to work with its counterparts to more rapidly develop advanced, reliable, low-cost lithium-ion batteries.
3. The Chinese universities and institutions have shown broad interest in cooperation. DOE should pursue such cooperation as a way to learn more about Chinese battery technology.
4. CITIC Guoan MGL; MGL New Energy Technology Co., Ltd.; Green Power Co.; Tianjin Lintian Double-Cycle Tech. Co.; Suzhou Phylion Battery Co., Ltd.; Tianjin Lishen Battery Joint-Stock Co., Ltd.; and Tongji University have expressed interest in having their staff do postdoctoral fellowships in the United States. This is an opportunity for national laboratories and the industry to build stronger relationships with Chinese companies and institutions.

BIBLIOGRAPHY

Publications

[1] Arai, E. B., et al. (2002). *Journal of Electrochemical Society* 149(4):A401–A406.
[2] Barnet, C., and A. O'Cull, 2003, "Embedded C Programming and Microchip P/C," Thomson Delmar Learning, John Wiley, New York, N.Y., U.S.A.
[3] Bitsche, O., Gutmann, G., Schniolz, A. & d'Ussel, L. (2001). "DaimlerChrysler EPIC I — Minivan Powered by Lithium-ion Batteries," in Proceedings of the Electric Vehicle Symposium, EVS- 18, Berlin, Germany.
[4] Black & Decker Press Release. (2005). Baltimore, Md., U.S.A., Nov.
[5] Broussely, M. (1999). "Recent Developments on Lithium Ion Batteries at SAFT," *Journal of Power Sources, 81–82*,140–143.
[6] Busch, R. & Schmitz, P. (2001). "The Ka, an Electric Vehicle as Technology Demonstrator," in Proceedings of the Electric Vehicle Symposium, EVS- 18, Berlin, Germany, Oct.
[7] Chan, C. C. (2005). *"The Challenge of Sustainable Mobility: Clean, Efficient and Intelligent Vehicles,"* company paper, Dalian, China.
[8] Cherry, C. (2006). *"Implications of Electric Bicycle Use in China: Analysis of Costs and Benefits,"* presented at the Future Urban Transport — Volvo Summer Workshop, held at the University of California, Berkeley Center, Berkeley, Calif., U.S.A.
[9] China Electric Vehicle Conference. (2005). China Electrochemical Society and China — Society of Automobile Engineers (CES & C-SAE), Wuhan, China, Sept. 2–6.
[10] China Hybrid Electric Vehicle Workshop. (2006). Beijing, China, July 14.
[11] Chung, C. H., et al. (2002). "Electronically Conductive Phospho-Olivines as Lithium Storage Electrodes," *Nature Materials, 1*,123-128.
[12] Dahn, B. Z., et al. (1994). "Lithium-Ion Battery Research," *Solid State Ionics, 69*, 265.
[13] Demuth, H. B. & Beale, M. (2004). *Neural Network Toolbox,* Mathworks, U. K.
[14] Economic Research Force of the China State Council Development Center. (2006). *Development Strategy of LEV in China Research Report,* Beijing, China.

[15] Edition of Selection of Achievements of 863 of China in Field of Energy. (2005). Sept. Engineering and Science and Technology Forum on Vehicle Energy Saving Technology and Development of New Energy Vehicles. (2005). Beijing, China, Oct.

[16] Gaines, L. & Cuenca, R. (2000). *"Costs of Lithium-Ion Batteries for Vehicles,"* unpublished report, Center for Transportation Research, Energy Systems Division, Argonne National Laboratory, Argonne, Ill., U.S.A.

[17] Guangyu, T., Li-an, Z., Xiaodong, H. & Quanshi, C. (2002), "Hybridization of Fuel Cell Power Train for City Bus," in Proceedings of the Electric Vehicle Symposium, EVS-19, Busan, South Korea.

[18] Hagan, M. T., Demuth, H. B. & Beale, M. (1996). *Neural Network Design,* MIT Press, Boston, Mass., U.S.A.

[19] "Hydrogen Fuel & Vehicle Standards Forum,". (2005). China Automotive Technology and Research Center, GM, China National Vehicle Standard Committee, Beijing, China, Nov. 4.

[20] Jamerson, F. & Benjamin, E. (2005). *Electric Bicycle World Report,* 7th ed., www.ebwr.com.

[21] Jian, X., Quanshi, C., Guangyu, T. & Hua, G. (2002). "Development of an Advanced Fuel Cell City Bus," in Proceedings of the Electric Vehicle Symposium, EVS-19, Busan, South Korea.

[22] Li, S. & Liqing, S. (2002). *"Research and Development of Electric Vehicle in China,"* Press of Beijing Institute of Technology, Beijing, China.

[23] Liqing, S., Sun, X. & Chen, J. (2004). *"Review of R&D and Market Predication of Hybrid Electric Vehicle for Green Transportation,"* in Proceedings of World Engineers Convention, Shanghai, China.

[24] Liqing, S., Sun, X., Wen, J., Chen, J. & Heliang, Z. (2004). *"Overview of the LV Research in China,"* IEE-VPP, CIBF 2005, China.

[25] Liqing, S., Sun, X., Yan, L., Yong, C. & Peiji, S. (2004). "Fuel Cell Progress and Its Application in Field of Transportation in China HFC," publication of the Beijing Institute of Technology, Beijing, China.

[26] Lithium Ion Battery Data Sheet. (2003). *Panasonic*, Tokyo, Japan.

[27] MacNeil, C. K., et al. (2000). *Journal of Electrochemical Society, 147(3)*, 970–979.

[28] Ober, J. (1999). *Lithium., U.S. Geological Statistics Yearbook,* United States Geological Survey, Information Circular No.8767, Washington, D.C., U.S.A.

[29] Picton, P. (2000). *Neural Networks,* Palgrave-Macmillon, New York, N.Y., U.S.A.

[30] Predko, M. (2000). *Programming and Customizing PICmicro Ivlicrontrollers,* McGraw-Hill, New York, N.Y., U.S.A.

[31] Proceedings of the 4th International Clean Vehicle Technology Conference. (2005). November 23– 25, Beijing.

[32] Proceedings of Symposium of Electric Vehicle, Clean Vehicle and Vehicle Environment Protection Technology. (2004). entire document, editor: W. Jiqiang, Shanghai, China.

[33] Quanshi, C., Jixin, X., Li, S., Yu, Z., Pingxing, X. & Liqing, S. (2005). "Research and Development of Electric Vehicle in China 2005," Press of Beijing Institute of Technology, Beijing, China, Sept.

[34] Rao, V. & Rao, H. (1995). *C++ Neural Network and Fuzzy Logic,* MIS Press, New York, N.Y., U.S.A.

[35] Ritchie, A. G. (2004). "Recent Developments and Likely Advances in Lithium Rechargeable Batteries," *Journal Power Source, 136*, 285–289.
[36] Sino, G. F. (2005). France Forum on Battery (RUM), by Transportation Department, France, and MOST of China, Bordeaux, France, April 6–9.
[37] St.-Pierre, C., Carignan, C., Pomerleau, D., St.-Germain, P. & Riddell, E. (2001). "Integration of the Lithium-Metal-Polymer Battery in the Ford TI-FINK City," in Electric Vehicle Symposium, EVS- 18, Berlin, Germany.
[38] Takeshita, Y. (2006). "Market Growth of Lithium-ion Batteries in Power Tool and Related Applications," The 2nd International Symposium on Large Lithium-Ion Battery Technology and Application (LLIBTA), Baltimore, Md., U.S.A.
[39] "Vehicle Fuel Saving Technology and New Energy Vehicle Engineering Forum,". (2005). Chinese Academy of Sciences, China, Society of Automotive Engineers, Beijing, China, Oct. 29.
[40] Wang, J. (2006). "Brief Overview of Chinese Battery Industry and Market," presented at the China International Battery Forum, Beijing, China.
[41] Weinert, J. & Ma, C. et al. (2006). "The Transition to Electric Bikes in China and Its Effect on Travel, Transit Use, and Safety," presented at the Transportation Research Board, Washington, D.C., U.S.A.
[42] Yu, J. S., Kim, S. W., Choo, H. S. & Kim, M. H. (2002). *"Recent Development of 8-A•h-Class High-Power Li-ion Polymer Battery for HEV Applications at LG Chem,"* Electric Vehicle Symposium 19 (EVS 19), Busan, South Korea.

Web Resources

The Administrative Commission of Wuhan East Lake High
Technology Development Zone
#450 Luoyu Rd. Wuhan
CN, Hubei 430079
China
www.chinaov.org

Alibaba.Com
6/F, Chuangye Mansion
East Software Park, 99 Huaxing Road
Hangzhou 310012, China
http://www.alibaba.com/companies/1009/Storage_Batteries_Secondary_Batteries.html

Battery University
Cadex Electronics
www.batteryuniversity.com

China Electric Vehicles
www.chinaev.com.cn

China Industry Development Report
http://www1.cei.gov.cn/ce/e_report/hy/ddc.htm

China Plugs into Electric Vehicles
Battery Breakthrough Holds Promise for Hybrids, Fuel Cells
By Miguel Llanos
MSNBC
http://msnbc.msn.com/id/6290392/
November 22, 2004
http://www.jr.co.il/articles/china-electric-vehicles

China's Electric/Hybrid Vehicle and Alternative Fuel Vehicle Programs
A July 1998 report from U.S. Embassy Beijing
http://www.usembassy-china.org.cn/sandt/Elecab.htm

Chinese Manufacturers Enter Lithium Battery Market
Advanced Battery Technology, Feb 2003
http://www.findarticles.com/p/articles/mi_qa3864/is_200302/ai_n9170533

www.Cleanauto.com.cn

Demonstration Project for Fuel Cell Bus
Commercialization in China
www.chinafcb.org

Electric Drive Transportation Association
1101 Vermont Ave. NW, Suite 401
Washington, D.C., 20005
www.electricdrive.org

EVS21.org
http://www.evs21.org/rubrique42.html
Frost and Sullivan Research Service
http://www.frost

Global Sources
c/o Media Data Systems Pte., Ltd.
Raffles City
PO Box 0203, Singapore 911707
http://www.globalsources.com/gsol/I/Lithium-ion-manufacturers/b/2000000003844/3000000181565/22761.htm

Global Sources
c/o Media Data Systems Pte., Ltd.
Raffles City
PO Box 0203, Singapore 911707
http://www.globalsources.com/manufacturers/Rechargeable-Lithium-Battery.html

GLOBALSPEC
Engineering Search Engine
Product Categories for polymer lithium battery
http://electronic-components.globalspec.com/Industrial-irectory/polymer

Green Car Congress
http://www.greencarcongress.com/2005/06/degussa_and_ena.html
and http://www.greencarcongress.com/electric_battery/index.html

Guangzhou Langging Electric Vehicle Co., Ltd.
Langqing Industrial Park No 1 Langqing Road
Taishi Village Dongchong Town, Panyu District
Guangzhou, P.R. of China
http://www.langqing.com

Hybrid Cars
http://www.hybridcars.com/lithium

Institute of Electrical Engineering
Academy of Sciences in the Czech Republic
Dolejskova 5, 182 02 Prague 8, Czech Republic
www.iee.cas.cz

Lithium Nanocomposite Polymer Electrolyte Batteries
Advanced Battery Technology, Jun 2002
http://www.findarticles.com/p/articles/mi_qa3864/is_200206/ai_n9088854

Market Research.Com
http://www.marketresearch.com/map/prod/974876.html

Market Research.Com
http://www.marketresearch.com/product/display.asp?productid=974876&xs=r&SID=807 12779- 350097775-269583 652&curr=USD&kw=&view=toc

MINDBRANCH
131 Ashland Street
North Adams, MA 01247
http://www.mindbranch.com/listing/product/R2-938.html

NERAC
One Technology Drive
Tolland, CT 06084
http://www.nerac.com/samples/lithium

People.Com.Cn
China Ushers in Electric Vehicles
http://english.people.com.cn/200207/31/eng20020731_100642.shtml

People's Daily Online
http://english.people.com.cn/200207/31/eng20020731_100642.shtml

Sanyo, GP Batteries Establish Lithium Battery JV
Advanced Battery Technology, Sep 2004
http://www.findarticles.com/p/articles/mi_qa3864/is_200409/ai_n9455849

SL Power Electronics™
Headquarters
6050 King Drive
Ventura, CA 93003
www.Slpower.com

Society of Automotive Engineers of China
www.sae-china.org

Source Guides
http://energy

Suzhou Phylion Battery Co., Ltd.
No. 81, Xiangyang Road,
Suzhou New District, Jiangsu 215011
www.xingheng.com.cn

Thunder Sky Energy Group Limited
Lisonglang Village, Gongming Town, Bao'an Dist., Shenzhen, P.R.C
Post code: 5181016
www.Thunder-sky.com

Tianjin Bluesky Double-Cycle Tech. Co., Ltd.
13, Road 4, Haitai Development, Huayuan
Industrial Park of Tianjin
New Tech Industrial Zone, 300384
www.tjshuanghuan.com

Tianjin Peace Bay Power Sources Co., Ltd.
No. 15 Kaihua Road, Huayuan Park
Hi-tech Industry Area, 300191
www.peacebay.com

Zhongyin (Ningbo) Battery Co., Ltd.
Ningbo Battery & Electrical Appliance I/E Co., Ltd.
99 Dahetou St., Duantang, Ningbo, China
http://www.sonluk.com/pro,lithium,round.htm.htm

APPENDIX A: ORGANIZATIONS REPRESENTED IN INTERVIEWS

Lithium-Ion Battery Companies and Institutions in China

1. Dr. Tian Guangyu, Professor
   Tsinghua University
   Department of Automotive Engineering
   Beijing, 100084

2. Dr. Lin Chengtao, Professor
   Tsinghua University
   Department of Automotive Engineering
   Beijing, 100084

3. Mr. Wei Feng, Manager
   Division of International Cooperation & Trade Promotion
   China Electrotechnical Society
   Beijing 100823

4. Dr. Sun, Li, Secretary General
   Chinese Electrotechnical Society Electric Vehicle Institution
   North China University of Technology
   Beijing 100041

5. Dr. Liqing Sun, Professor
   School of Mechanical Engineering & Vehicular Engineering
   Deputy Secretary General, Special Committee of Electric Vehicles
   China Electrotechnical Society
   Beijing 100081

6. Dr. Zhou Sigang, Professor
   China Electrotechnical Society
   Executive Deputy Secretary General
   Beijing 100823

7. Li Yongwei, Administrative Vice Director
   Research Institute
   CITIC Guoan MGL
   MGL New Energy Technology Co., Ltd.
   Beijing 102200

8. Wu Ningning, Vice Director
   Research Institute
   CITIC Guoan MGL
   MGL New Energy Technology Co., Ltd.
   Beijing 102200

9. Yuan Chun Huai
   Beijing Green Power Technology Co., Ltd.
   Beijing100086

10. Dr. Wang Haoran
    Beijing Green Power Technology Co., Ltd.
    Beijing100086

11. Dr. Zhang Bao Wen, Senior Researcher
    Beijing Green Power Technology Co., Ltd.
    Beijing100086

12. Dr. Wang Ji Qiang, Vice Chief Engineer and Professor
    Tianjin Institute of Power Sources
    Tianjin 300381

13. Ms. Yu Bing, Engineer/Object Manager
    Testing Center of Chemical & Physical Power Sources
    Ministry of Information Industry
    Testing Center of Electric Vehicle Battery of National 863 Project
    Tianjin 300381

14. Dr. Xu Gang, Vice-President
    Tianjin Lishen Battery Joint-Stock Co., Ltd.
    Tianjin 300384

15. Wang Guo Ji, Vice Manager
    Tianjin Lantian Double-Cycle Tech. Co., Ltd.
    Tianjin 300384

16. Zhao Chunming, Master Senior Engineer
    China Automotive Technology & Research Center
    Tianjin 300162

17. Dr. Wu Xiao Dong, Tech. Dept. Manager
    Suzhou Phylion Battery Co., Ltd.
    Jiangsu 215011

18. Dr. Zhang Lu, Tech. Dept. Testing Center
    Suzhou Phylion Battery Co., Ltd.
    Jiangsu 215011

19. Xuezhe Wei, Professor
    School of Automobile Engineering
    Tongji University
    Shanghai FCV Powertrain Co., Ltd.
    Shanghai 201804

20. Wei Yang
    School of Automobile Engineering
    Tongji University
    Shanghai FCV Powertrain Co., Ltd.
    Shanghai 201804

21. Dr. Zhang Xiangjun, Director
    General Research Institute for Nonferrous Metals
    Institute of Energy Materials and Technology Lab of Li-ion Battery
    Beijing 100088

22. Jin Wei Hua, Senior Engineer
    General Research Institute for Nonferrous Metals
    Division of Mineral Resources, Metallurgy & Materials
    Beijing 100088

23. Wu Guoliang, Vice General Manager
    Weifang Jade Bird Huanguang Battery Co., Ltd.
    Weifang, Shandong 261031

24. Dr. Zhang Jun, Professor
    Beijing Institute of Technology
    BTI EV Center of Engineering and Technology
    Beijing 100081

25. Dr. He Hong-Wen, Professor
    Beijing Institute of Technology
    BTI EV Center of Engineering and Technology
    Beijing 100081

26. Dr. Wang Zhen-po, Professor
    Beijing Institute of Technology
    BTI EV Center of Engineering and Technology
    Beijing 100081

27. Jinhua Zhang, Vice President
    China Automotive Technology & Research Center
    Beijing 100070

28. Hou Fushen, Director
    Hi-tech Development Department
    China Automotive Technology & Research Center
    Beijing 100070

APPENDIX B. LITHIUM-ION BATTERY TECHNOLOGY PRESENTATION

B.1. Current Status: HEV Batteries

- Conventional lithium-ion batteries for HEVs appear about ready for commercialization. U Major focus remains on cost reduction.
- Low-temperature performance and abuse tolerance still remain issues.
- Emerging technologies with nanostructure materials ($Li_4Ti_5O_{12}$ or $LiFePO_4$) appear to address these issues.
- Batteries, even those incorporating "stable" materials, will require appropriate thermal management controls and electronic protection circuits to extend battery life and avoid thermal runaway.
- Battery life projections of 10–15 years are based on limited data.

B.2. Lithium-Ion Battery Technology

Advantages	Disadvantages
Highest Energy Storage	Relatively Expensive
Light Weight	Electronic Protection Circuitry
No Memory Effect	Thermal Runaway Concern
Good Cycle Life	3-h Charge
High Energy Efficiency	Not Tolerant of Overcharge
High Unit Cell Voltage	Not Tolerant of Overcharge or Over- Discharge

B.3. Microsun's Lithium-Ion Battery

High-Power Lithium-Ion Battery Design

- Cell Specifications:
    - Commercially Available Type HPPC 18650 (high power)
    - Cutoff Voltage: 3.0 to 4.2 V
    - Cell Rated Capacity: 1.6 A·h
    - Maximum Discharge: 20 A
    - Maximum Charge: 5 A (80% in < 15 min)
    - Cycle Life: >1,000 (80% charge)
- Module Specifications:
    - Module Design: 4 series × 5 parallel
    - Nominal Voltage: 14.4 V
    - Module Capacity: 8 A·h
    - Maximum Discharge: 50 A continuous, 100 A of 10-s pulses

- Charge Time: < 1 h (80%)
- Fully integrated cell balancing, safety circuit, and thermal management
- Dimensions: 4.30 in. × 5.30 in. × 2.75 in.

B.4. Gold Peak Industries North America

18650 at 2,650 mA·h and 42.5 g – 230 Wh/kg, 0.67 amp/cell (3 parallel)
Both High-Capacity Versions

Parameter	LiSO$_2$ Primary	Lithium-Ion
A	2 0	67
A·h	8.25	2.65
V	2.72	3.67
Cells	1.00	1.00
Wh	22.44	9.73
Kg	0.0850	0.0421
Wh	264	231

B.5. A 123 26650 Lithium-Ion Specifications

Capacity	2.3 A·h
Energy	7.6 Wh (110 Wh/kg)
Nominal voltage	3.3 V
Cylindrical cell dimensions	25.9 mm dia, 65.4 mm H
Cell volume	34.45 cm3
Cell mass (without external tabs)	70 g
Impedance (1 kHz)	8 MΩ
Impedance (10 A, 10 s)	15 MΩ
Temperature range	−30°C to 60°C

B.6. Kokam America

- Fast charge capability: max. 3°C
- High discharge capability: 10 ~ 20°C
- High power density: over 1,800 W/kg (high power cell)
- Longer cycle life: over 2,500 cycles @80% DOD
- Wide operating temperature: -30 ~ 60°C
- Environmental friendly: zero emissions
- Low energy consumption: lightweight
- Maintenance-free operation
- Low heat emission in high-discharging mode

Parameter	Power Cell	Energy Cell
Energy Density		
Wh/kg	120	200
Wh/l	240	400
Power Density		
W/kg	2,400	550
W/l	4,800	900

B.7. Electro Energy, Mobile Products, Inc., Bi-Polar Lithium-Ion Battery Technology

- $LiCoO_2$ chemistry
- Cell capacity: 20 A•h
- Cell is capable of 5°C charge and discharge
- 8 stacks of cells, total number of cells is 112
- Total capacity: 160 A•h
- System energy: 8 kWh
- Cell weight: 100 lb
- Additional battery hardware weight (5–10 lb) will be required

B.8. Saft High-Power Lithium-Ion Cells (Vl20P)

- Nominal voltage: 3.6 V
- Average capacity: 20 A•h, 1°C after charge to 4.0 V/cell
- Minimum capacity: 18.5 A•h, 1°C after charge to 4.0 V/cell
- Specific energy: 187 Wh/kg
- Specific power: 1,811 W/kg
- Cell diameter: 41 mm
- Cell height: 145 mm
- Typical cell weight: 0.8 kg

B.9. High-Power Toyota 12-A·H Cell Lithium-Ion Battery

- Voltage: 3.6 V
- Capacity: 12 A•h
- Specific power: 2,250 W/kg
- Specific energy: 74 Wh/kg
- Weight: 580 g
- Dimensions: 120 mm (L), 25 mm (W), 120 mm (H)

B.10. Current Status: Technology Characteristics

Conventional lithium-ion technology
- Accurate SOC
- Excellent power density
- Good energy density
- Well matched for charge-sustaining

Emergent lithium ion technology (titanate anode or iron phosphate cathode)
- SOC determination problematic
- Good power density
- Very good energy density
- Well matched for PHEVs and potentially charge-sustaining HEVs

NiMH
- Difficult to ascertain the SOC accurately
- Good power density
- Abuse tolerant and proven technology
- Moderate energy density; *good for charge-sustaining HEVs*

	HEV (Cell Energy, 10-s Power)	**EV** (Cell Energy, 30-s Power at 80% DOD)
Conventional	70 Wh/kg 2,500 W/kg	140 Wh/kg 500 W/kg
Emergent	100 Wh/kg* 2,000 W/kg	
NiMH	50 Wh/kg 1,000 W/kg	65 Wh/kg 150 W/kg

* Projected

B.11. Challenges for Lithium-Ion Large Battery Development

- Abuse tolerance – material and battery management
- Cost: cathode selection, volume, standardization, packaging, battery management
- Life: cathode selection, operating temperature, packaging
- Performance in extreme temperatures — all aspects of chemistry

B.12. Lithium-Ion Battery Status vs. Goals for Power-Assist HEV

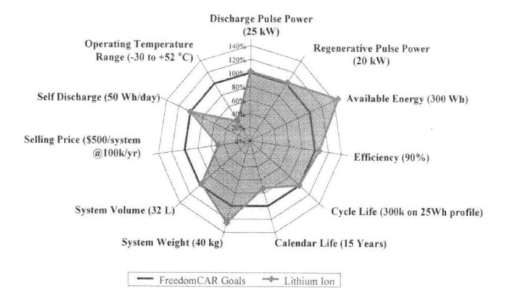

B.13. Electric and Hybrid Vehicle Battery Requirements (Module Basis)

Requirement Parameters	HEV	PHEV-20	PHEV-60
Vehicle ZEV Range (mi)	0	20	60
Battery Capacity (kWh)	<3	6	18
Cell Size (range corresponds to battery voltage 400V to 200V)	5–10	15–30	45–90
Specific Energy (Wh/kg)	>30	~50	~70
Specific Power (W/Kg)	~1000	~440	~390
Cycle Life			
Deep (80% DOD)	n.a.	>2,500	>1,500
Shallow (+/- 100 Wh)	200k	200k	200k

B.14. Battery Requirements for Transportation

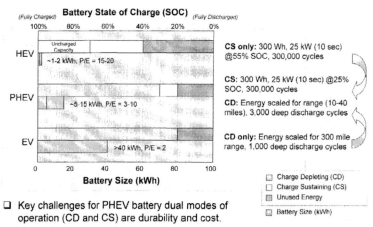

- Key challenges for PHEV battery dual modes of operation (CD and CS) are durability and cost.

B.15. Advanced Battery Technologies Development Status

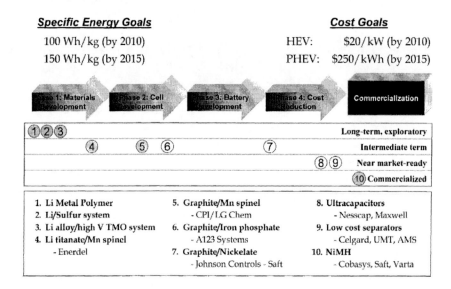

APPENDIX C. INTERVIEW QUESTIONS AND DISCUSSION TOPICS

The following questions were submitted to organizations developing lithium-ion batteries for discussions two weeks before meetings in the PRC.

1. Battery technology status: What is the level of development? Cell, module, or full pack?
2. Battery technology applications, power, energy, volume, and weight?
3. Testing methodology used? Charge/discharge limitations, cycle life, capacity, voltage, temperature operating range, and effect of aging?
4. Number of units produced per year, manufacturing facilities, and equipment used?
5. Any special feature(s) available on your current products?
6. Raw material availability domestically as well imported?
7. Where are your batteries sold? Domestically or exported? Are you working with any other companies overseas or domestically? How many batteries are produced for domestic use and for export purpose?
8. What is the cost of your battery domestically and overseas?
9. Is your battery cell, module, or pack available for testing and evaluation if we can work out a confidentiality agreement? Could we evaluate your technology at Argonne National Laboratory in the United States? When can you make the battery available for evaluation?
10. Are you conducting R&D at your facilities or with companies in China or overseas to improve your products or developing new products or technology? In general, what is the nature of your agreement? Is your company participating in joint venture or equity partnership?
11. Currently, are you working on batteries for electrical, hybrid, and/or plug-in hybrid vehicles?
12. Are you interested in pursuing battery development for electric, hybrid, and/or plug-in hybrid vehicles?
13. What is your company size? How many employees do you have? What are your sales per year (kW·h of capacity sold, or value sold in Yuan)?
14. Do you sell batteries directly as retail products? Do you sell batteries to other companies that convert them into packs with controllers? Is quality control causing you problems in selling batteries to some potential customers?
15. What are the incentives are offered by the Chinese Government to battery developers? Are these same incentives available to both domestic and/or international companies working with Chinese companies?
16. Do you have intellectual property, such as patents, joint venture agreements, or other rights to protect your products? How important are these for new product vs. improvement, venture vs. current manufacturer? Do you purchase battery technology and specialize only in production, or do you invest in battery R&D to develop your own products? If you do your own R&D, how much do you spend per year?

APPENDIX D. CHINESE LITHIUM-ION BATTERY DEVELOPMENT COMPANIES, CONTACTS, UNIVERSITIES, AND OTHER ORGANIZATIONS

List A

1. Beijing Waterwood Technologies Co., Ltd.
2. Topin Battery (China), Ltd.
3. Shenzhen Suppower Tech. Co., Ltd.
4. Shun Wo New Power Battery Technology, Ltd.
5. Sero Industrial & Commercial Co., Ltd.
6. Zhongyin (Ningbo) Battery Co., Ltd.
7. Shenzhen Win-top Electronic Tech. Co., Ltd.
8. Fujian Nanping Nanfu Battery Co., Ltd.
9. Hecell Hangzhou Battery Co., Ltd.
10. Shenzhen Mallerf Tech. Co., Ltd.
11. Hong Kong Eastar Industrial Co., Ltd.
12. Power Tech International Co., Ltd.
13. XELLEX Battery & Power Supply Tech. Co., Ltd.
14. Shenzhen Shunyi Industrial Co., Ltd.
15. GBP Battery Co., Ltd.
16. CLEScell International Battery Co., Ltd.
17. Wuhan Lixing (Torch) Power Sources Co., Ltd.
18. Sunhigh Battery(Primary Lithium) Co., Ltd.
19. Guangzhou Markyn Battery Co., Ltd.
20. Minamoto Battery Company
21. Golden Battery Technology Co., Ltd.
22. Wama Battery Co., Ltd.
23. Clescell International Battery Co., Ltd.
24. Able Battery Co., Ltd.
25. Guangzhou Great Power Battery Co., Ltd.
26. Narada Licom Power Tech. (Shanghai) Co., Ltd.
27. Megalink Company
28. Shenzhen Fekko Industrial Co., Ltd.
29. Totex International Limited
30. Hmc Power Technology, Ltd.
31. Tin Kam Company Limited
32. Shenzhen HJXY Li-Ion Battery Factory
33. Shenzhen Caixing Battery Factory
34. Hangzhou Lanbei Electric Bicycle Co., Ltd.
35. Jinhua Shiwei Vehicle Co., Ltd.
36. Hyper Power Co., Ltd.
37. Desay Power Technology Co., Ltd.
38. Yuntong Power Co., Ltd.
39. Huaye New-Technology Industry Co., Ltd.

40. Huangyuda
41. Huangyuda LTD Company
42. Zhongshan Mingji Battery Co., Ltd.
43. Hangzhou Huitong Industry Co., Ltd.

List B

1. Tianjin Lantian Double-Cycle Tech. Co., Ltd.
2. Tianjin Lishen Battery Joint-Stock Co., Ltd.
3. Beijing Green Power Tech Co., Ltd.
4. Beijing MGL
5. Shenzhen BAK Battery Co., Ltd.
6. Advanced Battery Technology, Inc., Ltd.
7. Hunan Corun Hitech Co., Ltd.
8. Degussa – ENAX (Anqui) Power Lion Co., Ltd.
9. Shenzhen Win – Top Electronic Tech. Co., Ltd.
10. Hecell Hangzhon Battery Co., Ltd.
11. Power Tech International Co., Ltd.
12. GBP Battery Co., Ltd.
13. CLEScell International Battery Co., Ltd.
14. Guangzhou Markyn Battery Co., Ltd.
15. Golden Battery Technology Co., Ltd.
16. Guangzhou Great Power Battery Co., Ltd.
17. Narada LiCom Power Tech (Shanghai) Co., Ltd.
18. Totex International, Ltd.
19. Hmc Power Technology, Ltd.
20. Shenzhen HJXY Li-Ion Battery Factory
21. Shenzhen Caixing Battery Factory
22. Hangzhou Lanbei Electric Bicycle Co., Ltd.
23. Jinhua Shiwei Vehicle Co., Ltd.
24. Hyper Power Co., Ltd.
25. Yuntong Power Co., Ltd.
26. Huangyuda
27. Wuhan Lixing (torch) Power Co., Ltd.

Institution and University Contacts

1. Heliang Zhou, China Electro-Chemical Society, Beijing
2. Wang Tiqiang, Battery Technology Institute
3. Chen Quanshi and Dr. Jiang Fachao, Tsinghua University
4. Dr. Prof. Sun Fengchun, Beijing Institute of Technology
5. Prof. C.C. Chan, University of Hong Kong

List C

1. Tianjin Lantian Double-Cycle Tech. Co., Ltd.
2. Tianjin Lishen Battery Joint-Stock Co., Ltd.
3. Beijing Green Power Tech. Co., Ltd.
4. CITIC Guoan MGL
5. Suzhou Phylion Battery Co., Ltd.
6. General Research Institute for Nonferrous Metals (GRINM), Institute of Energy Materials and Technology Laboratory of Li-ion Battery

Institutions and Universities

1. China Electrotechnical Society, Beijing
2. Tsinghua University
3. Tianjin Institute of Power Sources
4. Tongji University
5. Beijing Institute of Technology

In: Battery Technology Advancements in China
Editor: John S. Gaglione

ISBN: 978-1-61470-563-5
© 2012 Nova Science Publishers, Inc.

Chapter 2

ADVANCED BATTERY TECHNOLOGY FOR ELECTRIC TWO-WHEELERS IN THE PEOPLE'S REPUBLIC OF CHINA

Argonne National Laboratory and Energy Systems Division

NOTATION

Acronyms and Abbreviations

AC	alternating current
ASK	A-SI-KA Electric Bike Company, Inc.
ASME	American Society of Mechanical Engineers
B&K	Shenzhen B&K Technology Co., Ltd.
BAK	Shenzhen BAK Battery Co., Ltd.
BIT	Beijing Institute of Technology
BOV	battery-operated vehicle
BSEB	bicycle-style electric bike
BYD	BYD Battery Co., Ltd.
BYN	Shenzhen Bo Yi Neng Co., Ltd.
CAE	computer-aided engineering
CATARC	China Automotive Technology & Research Center
CCC	China Compulsory Certification
Cd	cadmium
CE	Conformité Européene
CHALCO	Aluminum Corporation of China Ltd.
Chinalco	Aluminum Corporation of China (holding company)
CO	carbon monoxide
CO2	carbon dioxide
CPSC	Consumer Product Safety Commission)
CPU	central processing unit
DC	direct current

DEC	diethyl carbonate
DLG	DLG Battery (Shanghai) Company, Ltd.
DMC	dimethyl carbonate
DOD	depth of discharge
DOE	U.S. Department of Energy
DOT	U.S. Department of Transportation
DVD	digital video disc
E-bicycle	electric bicycle (same as E-bike or ETW; can be bicycle or scooter)
E-bike	electric bike (same as E-bicycle or ETW; can be bicycle or scooter)
EC	ethylene carbonate
EMC	ethylmethyl carbonate
EPA	Environmental Protection Agency
EPAC	electric pedal-assisted bicycle
ES	electric scooter
ETW	electric two-wheeler (same as E-bike or E-bicycle; can be bicycle or scooter)
EU	European Union
Euro I, II, etc.	European Emission Standard Phase I, II, etc.
EV	electric vehicle
FCB	flexible printed circuit
FCEV	fuel cell electric vehicle
FPC	flexible print circuit
GDP	gross domestic product
GHG	greenhouse gas
GM	General Motors Corporation
HEV	hybrid electric vehicle
HR	U.S. House of Representatives (bill)
IEC	International Electrotechnical Commission
IEEE	Institute of Electrical and Electronics Engineers, Inc.
IPR	international property right
ISO	International Standards Organization
IT	information technology
ITRI	Industrial Technology Research Institute
JAMA	Japanese Manufacturers Association
JCS	Johnson Controls – Saft Advanced Power Solution
JIS	Japanese Industrial Standard(s)
LCD	liquid crystal display
LCM	liquid crystal module
LED	light-emitting diode
LEV	light electric vehicle
Li	lithium
LPG	liquefied petroleum gas
LPRCRTS	Law of the PRC on Road Traffic Safety
Mg	magnesium
MGL	CITIC Guoan Mengguli Corp.
MIC	Motorcycle Industry Council

Mn	manganese
Ni	nickel
Ni-MH	nickel-metal hydride
NO$_x$	nitrogen oxides
OECD	Organization for Economic Cooperation and Development
OEM	original equipment manufacturer
OVT	Office of Vehicle Technologies (in DOE)
PC	propene carbonate
PCM	power control module
PDA	personal digital assistant
PE	polyethylene
PHEV	plug-in hybrid electric vehicle
PLIB	polymer lithium-ion battery
PM	particulate matter
PNGV	Partnership for a New Generation of Vehicles
PP	polypropane
PRC	People's Republic of China (China)
PSHEV	parallel series hybrid electric vehicle
PWM	pulse with modulator
QC	quality control
QSR	Quality System Review
R&D	research and development
RMB	renminbi (means Chinese currency)
ROC	Republic of China (Taiwan)
ROHS	Restriction of Use of Hazardous Substances
SASAC	State-Owned Assets Supervision and Administration Commission
SGS	SGS S.A. or SGS Group (originally Société Genéralé de Surveillance; member of CE)
Si	silicon
SIAM	Society of Indian Automobile Manufacturers
Sn	selenium
SSEB	scooter-style electric bike
SUV	sport/utility vehicle
SVOC	semivolatile organic compound
TÜV	TÜV Rheinland Group
UL	Underwriters Laboratories Inc.
UPS	uninterrupted power supply
USABC	U.S. Advanced Battery Consortium
USD	U.S. dollar(s)
VC	vinylene carbonate
VOC	volatile organic compound
VRLA	valve-regulated lead acid
WEVC	Wanxiang Electrical Vehicle Co., Ltd.

Units of Measure

A	ampere(s)
A•h	ampere-hour(s)
C	hourly capacity rating (where 1 C = 1 hour) measured in A•h
°C	degree(s) Celsius
cc	cubic centimeter(s)
cm	centimeter(s)
dB	decibel(s)
ft^2	square foot (feet)
g	gram(s)
G	force from gravity
h	hour(s)
Hz	hertz
in.	inch(es)
kA	kiloampere(s)
kg	kilogram(s)
km	kilometer(s)
kW	kilowatt(s)
kWh	kilowatt-hour(s)
L	liter(s)
lb	pound(s)
m2	square meter(s)
mA	milliampere(s)
mAh	milliamp-hour(s)
mbd	million barrels per day
mi	mile(s)
min	minute(s)
mm	millimeter(s)
MMT	million metric ton(s)
mph	mile(s) per hour
nm	nanometer(s)
ppm	part(s) per million
s	second(s)
TOE	ton(s) of oil equivalent
V	volt(s)
W	watt(s)
Wh	watt-hour(s)
$	dollar(s)
¥	yuan(s)
mΩ	impedance

ACKNOWLEDGMENTS

My heartfelt thanks go to my colleague Dr. Yang Jianhong, who helped me plan my visit to China and who, while he was on vacation there, attended some meetings and made the arrangements for my trips to Beijing and Shenzhen. I also want to thank Dr. Wei Xuezhe, Head of the Advanced Automotive Laboratory at Tongiji University, who arranged for my trip to Shanghai.

I would like to thank Tien Duong, Office of Solar Energy in the U.S. Department of Energy's Office of Energy Efficiency and Renewable Energy (DOE-EERE), who initiated this project, for his steadfast support. My thanks also go to the current sponsor, David Howell, Team Leader, Hybrid and Electric Systems, Vehicles Technologies Program, DOE-EERE, for his continuing encouragement for this study.

I very much appreciate the efforts of Larry R. Johnson, Director of the Transportation Technology Research and Development Center at Argonne National Laboratory, and also of my Team Leader there, Danilo J. Santini, for guiding me in conducting the study and helping to make my visit a success.

Finally, I thank Marita Moniger of Argonne's Technical Services Division (TSD) for editing this report and Linda Graf and Lorenza Salinas (TSD) for formatting it.

SUMMARY

This report focuses on lithium-ion (Li-ion) battery technology applications for two- and possibly three-wheeled vehicles. The author of this report visited the People's Republic of China (PRC or China) to assess the status of Li-ion battery technology there and to analyze Chinese policies, regulations, and incentives for using this technology and for using two- and three- wheeled vehicles. Another objective was to determine if the Li-ion batteries produced in China were available for benchmarking in the United States.

The United States continues to lead the world in Li-ion technology research and development (R&D). Its strong R&D program is funded by the U.S. Department of Energy and other federal agencies, such as the National Institute of Standards and Technology and the U.S. Department of Defense. In Asia, too, developed countries like China, Korea, and Japan are commercializing and producing this technology. In China, more than 120 companies are involved in producing Li-ion batteries. There are more than 139 manufacturers of electric bicycles (also referred to as E-bicycles, electric bikes or E-bikes, and electric two-wheelers or ETWs in this report) and several hundred suppliers. Most E-bikes use lead acid batteries, but there is a push toward using Li-ion battery technology for two- and three-wheeled applications.

Highlights and conclusions from this visit are provided in this report and summarized here.

- In 2006, 20 million E-bikes were made in China. At present, China has 50 million battery-operated bicycles on the road, of which a very small percentage operate on Li-ion batteries. The rest of them use lead acid batteries. In China, about 2,500 companies produce electric two- or three- wheeled vehicles. All of the large

- companies producing electric vehicles (EVs) have E-bike models that are powered by Li-ion batteries, but the performance-to-price ratio for those E-bikes is still not compatible with that for E-bikes powered by lead acid batteries. This is the key reason that bikes powered by Li-ion batteries are still not in mass production.
- Energy is playing a key role in China's rapid development. Industrialization and the growth of the country's gross domestic product (GDP) depend heavily on the availability of affordable and reliable energy. The transportation sector depends on such energy as well. It appears that as people's incomes rise, they begin to travel farther and more often. With the recent rise in per-capita income in China, more people are able to afford cars and want the personal benefits that automobile ownership provides. As the automobile fleet grows, the demand for the fuels that the cars use and the supporting supply and distribution infrastructure for those fuels also increase.
- ETWs are a category of vehicles in China that include two-wheeled bikes that are propelled by human pedaling that is supplemented by electrical power from a storage battery (bicycle-style electric bikes or BSEBs) and low-speed scooters that are propelled almost solely by electricity (scooter-style electric bikes or SSEBs). Most riders of ETWs rely exclusively on electric power, not human pedaling. In most cities, E-bikes are allowed to operate in the bicycle lane and are considered to be a bicycle from a regulatory perspective (i.e., no helmet and no driver's license are required to operate one). The technology used by both types of ETWs is similar. BSEBs typically have 36-V batteries and 1 80–250-W motors. SSEBs typically have larger 48-V batteries and higher-power 350–500-W motors. They look more like motorcycles than bicycles. Regulations restrict E-bikes from going any faster than 20 km/h, but many of them, especially scooters, can go faster than that limit. In fact, some are advertised to go 40 km/h.
- Traffic safety is perhaps the most important issue associated with ETW growth. In November 2006, Guangzhou became the third city in China (after Fuzhou and Zhuhai) to ban ETWs, in response to advice from the traffic management bureau citing traffic safety concerns.
- The majority of the world's ETWs (96%) are concentrated in China. The next-largest ETW market is Japan, with annual sales of 270,000 bikes in 2006 and a 13% average annual growth rate since 2000. Pedelecs (a style of ETWs driven primarily by human power with battery assistance) are the dominant type of ETW. Most pedelec ETWs use nickel-metal hydride (Ni-MH) or Li-ion batteries. The battery capacity ranges from 0.2 to 0.6 kWh, the motor power ranges from 150 to 250 W, and the pedelec prices range from $700 to $2,000. In Europe, the market was estimated to be 190,000 bikes per year in 2006. Electric bikes in Europe are also mainly pedelecs. Sales are greatest in the Netherlands because of its extensive bicycle infrastructure and deep-rooted biking culture. Germany and Belgium are the next-largest markets for pedelecs.
- Electric bicycles have become popular in many Chinese cities because they provide several benefits to riders. They offer personal mobility, are easily accessible, save time, and are low in cost, and they are also environmentally friendly.
- In 2006, China produced 19.5 million E-bicycles.

- There are 10,000 enterprises, both large and small, involved in the Chinese national production of electric bikes. Small and mid-sized companies accounted for 35% of total national bike production in 2007. Most of the E-bikes use lead acid batteries, yet in 2007, the entire industrial production of Li-ion batteries for electric bicycles had surpassed 100,000 ETWs. In 2007, China exported about 395,000 electric bicycles; exports to Japan, the United States, and the European Union (EU) numbered 203,300, which was 58% of production. The number of electric bicycles that Japan, the United States, Italy, Holland, Germany, Hungary, and Great Britain imported from China accounted for 87,800 ETWs.
- The need for motorized personal transportation in China is increasing as its cities sprawl. Electric bicycles are an attractive option for commuters, service people, and couriers. At a cost of 1,500–3,000 ¥ (yuan), or U.S. $180–360, an electric bike is much more affordable than an automobile; its cost is only a small fraction of that of a car. Riding an electric bike is also exhilarating to some people. Riders simply hop on and crank the throttle, and an electric motor built into the hub can propel the bike to speeds of 20 km/h or more. Despite the appeal of electric bikes, some Chinese cities have banned them altogether because of concerns about environmental problems and public safety. But the bans have not stopped millions of Chinese people from buying ETWs. This development is astonishing to ETW advocates, who have been struggling for a decade to build a market for electric bikes in the United States and Europe.
- Crane Company, which produces 50,000 bikes a year and has a workforce of 210, is one of the few businesses that can sustain R&D. However, because of China's weak protection of intellectual property, the innovations made by companies like Crane spread quickly to the entire industry. Crane Company believes that more R&D is necessary to improve its products. Improvements in bike technologies — such as brushless motors that deliver higher torque, electronic controllers, and lead acid batteries that deliver a range of up to 60 km and last up to two years — are needed.
- According to the PRC's *Report on the Environment for 2005*, ownership of automobiles exceeded 43 million and ownership of motorcycles exceeded 94 million by the end of 2005. Compared to 2004 figures, the number of automobiles had increased by 20.6%, while the number of motorcycles had increased by 23.6%, according to the Chinese Federal Government's Environmental Protection Administration. Private ownership of cars showed a high annual rate of increase (23%), and the number of private vehicles in the PRC was 14.8 million, or about 55% of the total number of vehicles, in 2004.
- The PRC has adopted a roadmap for new vehicle standards, which lays out a schedule for introducing vehicle emission standards equivalent to the European emission standards for light-duty vehicles. The State Council of the PRC approved the implementation of Euro III (European Emissions Standard Phase III) in 2005 and Euro IV in 2007 in Beijing for light-duty and heavy-duty vehicles. The State Council required Beijing to ascertain the availability of fuel of the necessary quality by the time of implementation.
- According to the State Statistical Bureau, China produced 65,497,775 bicycles in 2007, representing a growth rate of 5.12% when compared with the same period in the previous year.

- Several countries around the world are involved in ETW R&D and production. These companies are targeting developments for Europe, Asia, China, Taiwan, India, and Japan. ETW vehicles are considered to be vehicles for the common people, allowing them to save energy, help reduce air pollution, and improve their lifestyles. These vehicles are the lowest-cost option available to the masses. This mode of transportation is a reasonable catalyst in finding socially, financially, and environmentally sound solutions to problems related to urban mobility.
- The number of electric bikes in the United States is small. The amount of product information that is available is very limited. Most major bicycle companies experimented with low power (250-W) electric bikes in the late 1990s. Consumers were disappointed with the products. Companies such as GT/Charger, Schwinn/Currie, Trek/Yamaha, Brunswick, ZAP, Ford, and Total EV/Merida offered electric bicycles in the late 1 990s. Most units were too expensive and were not powerful, and they were often not reliable.
- Electric bicycles are commonplace in Japan and other parts of Asia. Panasonic is one of Japan's largest manufacturers of electric bikes. The Panasonic electric bike is lightweight, foldable, comfortable, and it can easily climb any hill. It is light because of its Li-ion battery and advanced motor and controller system.
- Europeans are expected to purchase about 187,000 electric bikes in 2009. Most will be European brands that employ products or components from Taiwan, Japan, and China. It is estimated that the number of electric bikes in Europe will grow by more than 1 million every year. In more developed countries (like the Netherlands, Germany, and Switzerland), Europeans are willing to pay high prices for electric bikes that offer high quality and good performance. Other countries, such as Italy and those in Eastern Europe, need low-priced vehicles.
- Two- and three-wheeled vehicles are seen as the most potent vehicle options for zero local emissions in the near future. In India, local companies are developing these vehicles: An electric auto-rickshaw is undergoing user trials, and an ETW is in the prototype stage. So far, there has not been much business in E-bikes in India. However, E-bikes would be very appropriate for local Indian markets and should become popular as they become more available and as the people's incomes improve. Their price will be an important issue. It is predicted that eventually, the largest market after China will be India.
- Taiwan was the first country in the world to implement the zero-emission two-wheeler vehicle mandate. To support the government policy, the Industrial Technology Research Institute (ITRI) developed two generations of electric scooters by first implementing valve-regulated lead acid (VRLA) batteries and then implementing Ni-MH batteries.
- The demand for Li-ion rechargeable batteries has been driven by the rapid growth of electronic portable equipment, such as cellular phones, laptops, and digital cameras. Also contributing to their development in China and the world has been the expectation that rechargeable batteries will play a large role in alternative energy technologies — as well as in E-bikes, EVs, hybrid vehicles, and plug-in hybrid electric vehicles (PHEVs).

- Safety is a primary concern of Chinese government officials. In each of the three years 2004, 2005, and 2006, there were more than 100,000 road fatalities in China, and most of the victims were vulnerable road users, such as pedestrians or bicyclists (National Bureau of Statistics 2007). One motivation cited for regulating the use of gasoline-powered motorcycles is safety. Beijing officials cited safety as a main reason to ban E-bikes as well. The China Bicycle Association (E-bike advocates) countered, stating that the crash rate (percent of vehicles involved in a crash per year) for E-bikes is only 0.17%, while it is 1.6% for cars.
- The PRC is the largest populist country in the world. It has been involved in developing and manufacturing battery technologies for several years. Over the last 10 years, it has exported the most batteries for telecommunications, computers, cell phones, and other electronic equipment to many countries. Several hundred companies, small and large, are involved in developing lead acid, Ni-MH, and Li-ion batteries for these applications.
- Japan is the largest producer of Li-ion batteries in the world and owns most of the patents related to them. The Li-ion battery industry in China started later but developed very rapidly. On the basis of incomplete statistics, there are about 200 companies in China producing Li-ion batteries and related materials. It is predicted that China will soon be the biggest producer of Li-ion batteries, overtaking Japan.
- In recent years, BYD Battery Co., Ltd. (BYD), Tiajin Lishen Battery Joint-Stock Co., Ltd. (Lishen), and other Li-ion battery companies have been growing very rapidly. In 2006, exports of Li-ion secondary batteries numbered more than 1 billion cells, at a value of more than U.S. $2.98 billion, and the exports increased annually by 34% and 29%, respectively, for BYD and Lishen. Sony exports amounted to U.S. $256 million. Only BYD and Lishen are in the top-10 Chinese-owned exporting companies. Japanese, Korean, and Taiwanese ventures are still the main exporters of Li-ion batteries, accounting for 60% of the exports. In 2007, exports were expected to increase by 25%.
- Standards for the Li-ion battery industry need to balance product safety and performance. So far, various standards from China and abroad are used for testing. Tests cover short circuits, overcharging, overdischarging, vibration, punching, pressing, dropping, the heating box, a low-pressure atmosphere, temperature cycles, and other parameters to simulate normal and abnormal situations for battery applications. The test objectives are to develop good operating criteria and achieve ease of operation. The standards also provide a pathway for designing safe battery technologies with acceptable performance. The Chinese safety standards still need to be developed at an international level.
- As hybrid automobile technology is maturing, the power battery market for EVs is expanding. There are four "bike cities" in China: Beijing, Tianjin, Shanghai, and Chengdu. There are 10.5 million bikes in Beijing, 9.7 million in Tianjin, 9.2 million in Shanghai, and 7.5 million in Chengdu. A survey conducted by Tongji University in Shanghai indicates that as many as 76% of the citizens in big cities would like to use E-bikes instead of bikes, which means 350 million of the 450 million original bike customers would like to have E-bikes.

- In Shenzhen, it is said that more than 150 companies are making secondary batteries, including Li-ion, Ni-MH, lead acid, and nickel-cadmium (Ni-Cd) batteries. Of these companies, 95% are privately owned, and half are working on Li-ion batteries.
- CITIC Guoan Mengguli Corp. (MGL) is China's largest manufacturer of the Li-ion cathode material, LiCoO2, and it is the first in line to market the new cathode materials LiMn2O4 and $LiCo_{0.2}Ni_{0.8}O_2$. Being quality-oriented, MGL has been certified to both the New and Hi-Tech Enterprise standards and International Standards Organization Standard ISO 9001:2000. MGL's unique synthesis method simply and efficiently produces cathode materials of superior electrochemical performance and reliability in an environmentally friendly way.
- An investigation by the Electric Vehicle Institution in the Chinese Electrotechnical Society showed that in 2006, 20 million electric bikes had been made in China. There are now 50 million battery-operated bicycles on the road in China.
- BYD is the third largest manufacturer of rechargeable batteries and a world leader in Ni-MH, Ni-Cd, Li-ion, and lead acid cells and chargers. These products have a wide range of applications in power tools, toys, digital cameras, mobile phones, cordless phones, and other devices. BYD offers good-quality products at competitive prices.
- Shenzhen BAK Battery Co., Ltd. (BAK), produces 600,000 cells per day for cell phones, 150,000 cells (18650 type) per day for notebooks, and 20,000 polymer Li-ion battery cells per day for electric vehicles and electric bikes. Li-ion power batteries for E-bikes are still in the research stage; these batteries use four 2.5-A·h cells (26650 type cells) in parallel and then 11 cells in series to make a 10-A·h, 36-V battery pack. The range is 45–50 km per charge. BAK has patents for protective boards for the Li-ion battery pack. The positive material is $LiFePO_4$.

On the whole, the PRC is making significant progress in manufacturing Li-ion battery technologies and in developing and manufacturing E-bikes. The government has a national program in place to attract foreign companies to set up joint ventures and/or partnerships with Chinese companies. The Chinese government offers large incentives to Chinese companies that produce batteries for export. The Chinese government also gives Chinese-owned companies additional incentives to conduct research and provides capital for manufacturing Li-ion batteries for all applications.

1. INTRODUCTION

The sponsor of this report prepared by Argonne National Laboratory is the Office of Vehicle Technologies (OVT) in the U.S. Department of Energy (DOE) Office of Energy Efficiency and Renewable Energy. The report was prepared for the Hybrid and Electric Systems Team. The information in the report is based on the author's visits to the cities of Beijing, Tianjin, Shenzhen, and Shanghai in the People's Republic of China (PRC or China). In China, he met with representatives from organizations that are developing and manufacturing lithium-ion (Li-ion) battery technologies for cell phones, electronics, electric bikes, and electric vehicle (EV) and hybrid vehicle applications. (These organizations are listed in Appendix A.) The report focuses on Li-ion battery technology applications for two-wheeled and possibly three-wheeled vehicles. The purpose of the visits was to assess the status of Li-ion battery technology in China and assess Chinese policies, regulations, and incentives related to the technology and two- and three-wheeled vehicles. It was also to determine if Li-ion batteries produced in China would be available for benchmarking in the United States. Benchmarking would help DOE and U.S. industries that develop batteries better understand the status of the technology and formulate a long-term research and development (R&D) program. This report also provides information on joint ventures and on PRC government incentives and policies for joint ventures and cooperative programs.

1.1. Purpose of and Background on Report

The purpose of this report is to assess the status of Li-ion battery technologies in China in terms of the following parameters:

- Performance and life requirements of Li-ion batteries,
- Safety,
- International market for Li-ion batteries,
- Growth in the Li-ion battery industry,
- Growth in the two-wheeler market,
- Standards for Li-ion battery technology for two-wheelers,
- Advantages of and challenges facing Li-ion batteries for two-wheeler applications and key materials requirements, and
- Availability of two-wheeler applications.

In addition, information on joint ventures, partnerships, technology challenges, and methods of recycling batteries are described in this report.

Some U.S. and Chinese companies have begun to develop joint relationships to conduct R&D on and to manufacture advanced vehicle technologies. The information in this report on Chinese Li-ion battery technologies could help U.S. companies focus on technologies for which the United States would have a competitive advantage. By helping U.S. scientists and engineers select the best Chinese batteries for possible follow-up benchmarking and testing, this study could strengthen both U.S. and Chinese programs by helping them avoid undesirable duplication of effort and focus on their individual strengths in developing and manufacturing Li-ion batteries.

DOE's OVT program always strives to determine the best technologies that could be used by U.S. automobile manufacturers and to build the ability of domestic manufacturers to produce competitive or superior technologies. OVT's Energy Storage Program — through its testing and benchmarking of Japanese batteries — has provided U.S. manufacturers and customers with a degree of assurance about the reliability of battery technologies. In the process, the Energy Storage Program has given domestic manufacturers some of the information they have needed in deciding whether to proceed with programs to develop hybrid electric vehicles (HEVs) and plug-in hybrid electric vehicles (PHEVs). Greater confidence in the ultimate marketability of hybrid powertrain technologies will help ensure the success of OVT's vehicle development program.

The purpose of this study is consistent with past efforts. The U.S. Advanced Battery Consortium (USABC) and FreedomCAR Partnership for developing battery technologies will benefit from the results of this study. This study provides a Li-ion battery technology assessment that is specific to China and a list of contacts in key organizations and manufacturers. As DOE is embarking on a new "Plug-In Hybrid Initiative," this study identifies Li-ion battery technologies compatible with powertrain technologies being developed in the United States, thereby increasing the likelihood that such powertrains will succeed in reducing the rate of growth of world oil consumption.

In 2006, 20 million electric bikes were made in China. (In this report, the terms electric bike or E-bike, electric bicycle or E-bicycle, and electric two-wheeler or ETW are used interchangeably.) At present, China has 50 million battery-operated bicycles on the road. Only a very small percentage operate on Li-ion batteries; the rest of them use lead acid batteries. In China, about 2,500 companies produce ETWs or electric three-wheelers. All of the large companies producing EVs have models of electric bikes powered by Li-ion batteries, but the ratio of performance-to-price for these bikes is still not compatible with that for bikes powered by lead-acid batteries. This is the key reason that bikes powered by Li-ion batteries are still not in mass production. The price of a Li-ion battery pack is three to four times higher than the price of lead acid batteries. The increase in the price for a ton of lead — from 5,000 ¥ (yuan, which is interchangeable with RMB or renminbi, the Chinese currency) in 2006 to 25,000–34,000 ¥ in 2007 — was expected to enhance the market for two- or three-wheelers powered by a Li-ion battery pack; however, this result did not materialize.

The author visited battery research and testing organizations, battery technology societies, universities, electric bike developers, and organizations that were conducting R&D on Li-ion battery technology or manufacturing Li-ion batteries in the PRC (Appendix A). During the visits, he conducted 29 interviews. He made a presentation to each of these organizations on the status of Li-ion battery technology for EV, HEV, and PHEV applications in the United States (Appendix B). The people interviewed were in industry, government laboratories, and academia. Individuals from industry included representatives from material suppliers and battery manufacturers who served in technology development, management, manufacturing, and marketing positions. Each interviewee received a list of questions in advance that served as a guide to the interview process. Appendix D lists the questions used to guide the personal interviews. Interviews did not always follow the sequence of the listed questions. The interviews were conducted in a relaxed, conversational manner, allowing the experts to focus on what they considered to be the most important factors that influenced the development of Li-ion battery technologies and the decisions made by manufacturers about the production of Li-ion batteries. Responses from those who were interviewed helped Argonne identify and analyze developments in Li-ion battery technology in the PRC and obtain information about estimated costs and manufacturing capabilities.

Contacts in the United States were made by attending advanced battery technology meetings, seminars, and conferences, such as the Advanced Automotive Battery Conferences in 2007 and 2008. These contacts and conferences provided information on developments in Li-ion battery technology in the United States, Europe, and Asia. The conferences were extremely useful for gathering information to help understand the status of Li-ion batteries in these countries and how activities there compare with activities in the PRC.

1.2. Energy Situation in China

Energy is playing a key role in the rapid development of China. The country's industrialization and the growth in its gross domestic product (GDP) depend heavily on the availability of affordable and reliable energy. The transportation sector depends on such energy, too. In general, as people's incomes rise, they seem to travel farther. With the recent rise in per- capita income in China, more people have been able to afford cars and are wanting the personal benefits that automobile ownership provides. As the automobile fleet grows, the demand for the fuels the vehicles need to run and for the supply and distribution infrastructure that support those fuels will also increase.

Today, the Chinese use much less energy per capita than citizens of the member countries of the Organization for Economic Cooperation and Development (OECD). The average Chinese citizen, at 0.6 ton of oil equivalent (TOE) per capita per year, uses about 8% of the energy consumed by the average U.S. citizen and about 15% of the energy used by the average citizen of Japan or Germany. The high U.S. energy consumption is linked in part to the country's greater use of energy for transportation, which is, in turn, linked to its lower population density. Globally, an average of 1.4 TOE per capita per year is consumed. The challenge, then, is how to provide a source of inexpensive energy to developing countries as they seek to advance yet remain mindful of concerns about excessive dependency on oil imports and the need to limit global greenhouse gas (GHG) emissions.

From a strategic point of view, a shortage of domestic oil is a barrier to the development of an automotive industry. Motor vehicles in China consume 85% of the country's gasoline output and 42% of its diesel output. In 1995, China's demand for oil was 3 million barrels per day (mbd) or 147 million metric tons (MMT) per year; this demand grew to 4.5 mbd (220 MMT) in 2000 and was projected to reach 5.2 mbd (250 MMT) in 2005. In 2000, imports of petroleum were 70 MMT, and an annual increase in imports of at least 10 MMT per year was anticipated for the short term. According to predictions, by 2010, China will need 270–3 10 MMT of crude oil per year. Unfortunately, the domestic supply will reach just 165–200 MMT per year, and the deficit of 105–110 MMT will have to be imported.

The rapid growth of the vehicle sector is the primary force driving China's rapid shift from being a net petroleum-exporting country to a net importer. This shift not only creates concerns about China's energy security and balance of payments but also increasingly strains China's refinery sector. China's refineries have traditionally been largely able to provide the country's own refined product needs by using a refining network set up for indigenous heavy, sweet crudes. A particular concern is the high sulfur content of imported crude oil when compared with that of domestic crude. Because the Chinese refineries were built to process the relatively low-sulfur domestic crude, their available hydro-desulfurization capacity is limited.

The quality of fuels is inextricably linked to regulations on vehicle emissions performance. China has decided to follow the pollution control strategies of the European Union (EU), and it will upgrade its fuel quality, including further reducing the fuel's sulfur content, to meet those emission standards. The fuel specification standards that are now in effect in the EU will be required for fuels sold in 2012. The European Union Commission initially proposed requiring the introduction of gasoline and diesel fuels with a sulfur content of less than 10 parts per million (ppm) or 0.00 1% by mass as early as 2005, with a complete shift to these low-sulfur fuels by 2011. Because lower sulfur levels in gasoline and diesel fuel are preconditions for the introduction of advanced vehicle technologies that are able to comply with future European Emission Standard III (Euro III) and Euro IV standards, China will have to substantially upgrade its refineries.

In addition, the fuel efficiency of most Chinese cars today is worse than that of cars of comparable weight and size in industrialized countries. Unless fuel economy is improved in the future, even greater strains will be placed on the refinery sector. China is anticipating, at a minimum, a threefold increase in its vehicle fleet, not including motorcycles, between 2002 and 2020. The automobile fleet, in particular, is expected to increase by a factor of between four and five within the same time period. On the basis of the vehicle characteristics in the 10th five-year plan, it is estimated that total fuel consumption will more than double by 2020 despite a gradual improvement in the fuel efficiency of gasoline vehicles and an increase in the use of more efficient diesel technology.

The overall implication of this prediction is that although improvements in vehicle efficiency will help reduce fuel demand as the Chinese car fleet expands over the coming decades, the improvements will not offset the increased use of petroleum. Use of smaller vehicles with lower average fuel consumption could reduce fuel consumption. Such use will promote the development of electric two wheelers (bikes and motor bikes), which may be an attractive option to Chinese customers.

1.3. Description and Benefits of Electric Two-Wheelers

There are many different types and sizes of two-wheelers around the world. Table 1-1 classifies the types of two-wheelers most commonly used in China according to their key attributes.

Electric two wheelers (ETWs) are a category of vehicles that includes (1) two-wheeled bikes propelled by human pedaling and supplemented by electrical power from a storage battery (bicycle-style electric bicycles or BSEBs) and (2) low-speed scooters propelled almost solely by electricity (scooter style electric bicycles or SSEBs). Other names for ETWs are electric bicycles or E-bicycles and electric bikes or E-bikes. Most riders of ETWs rely exclusively on electric power, not human pedaling. In most cities, electric bikes are allowed to operate in the bicycle lane and are considered bicycles from a regulatory perspective (i.e., riders do not need a helmet or driver's license to operate them). The technology for each type of ETW is similar. The main components of an ETW include a hub motor, controller, and battery. BSEBs typically have 36-V batteries and 1 80–250-W motors. SSEBs typically have larger 48-V batteries and higher- powered 350–500-W motors. Regulations limit electric bikes to a maximum speed of 20 km/h, although many of them, especially scooters, can travel at higher speeds; some are advertised to go 40 km/h. They can travel at speeds from 25 to 40 km/h at a range of 25 to 50 km on a single charge, which requires 6 to 8 hours. Because batteries for electric bikes are recharged from a standard electrical outlet, they require no new infrastructure. The majority of ETW users recharge their bikes at home during the night when electricity is cheaper. In urban areas, this practice typically means that either the battery or the entire ETW is carried into a multilevel apartment building. It is also common to see bikes being charged during the day outside ground- floor shops by using standard electrical outlets.

ETWs have become a popular transportation mode for Chinese consumers because they provide a convenient yet relatively inexpensive form of private mobility. Thus, they are an attractive alternative to public transit or regular bicycling. Figure 1-1 compares the cost (in cents/km in U.S. dollars or USD) and in-use speed of ETWs versus other modes of transport.

Figure 1-1 shows that the key cost advantage of ETWs over motorcycles is their lower operating cost (which results both from using a less-expensive fuel and from using it more efficiently), even after accounting for battery replacement cost. While the initial cost of ETWs is also lower than that of motorcycles, motorcycles presumably have a longer lifetime; thus, the levelized vehicle purchase costs are roughly equal. Not surprisingly, ETWs are faster than bicycles. Speeds of motorcycles (scooters) that run on liquefied petroleum gas (LPG) are even higher in free-flow conditions. All modes approach the same speed when flow is congested.

ETWs are promoted by national and many local governments because of their low energy consumption and zero tailpipe emissions, benefits that are especially important in China's congested urban areas. In recent years, however, a handful of cities have decided to ban electric bikes, stating reasons related to decreased safety and traffic flow efficiency that occur when they are mixed together with engine-powered cars and trucks. Cities like Guangzhou have banned all motorized two-wheelers in favor of public transportation, bicycles, and cars. Some cities choose to neither support nor ban them. The use of E2Ws as an urban travel mode has both positive and negative attributes; the main ones are listed in Table 1-2.

Table 1-1. Classification of Chinese Two-Wheelers.

Class	Type	Power (Engine Size in kW)	Top Speed (km/h)	Energy or Fuel Use per 100 km	Range (km)	Picture
Bicycle		Not applicable	10–15	Not applicable	Not applicable	
Electric two-wheeler	Electric bicycle (BSES)	0.25–0.35	20–30	1.2–1.5 kWh	30–40	
	Electric scooter (SSEB)	0.3–0.5	30–40	1.5 kWh	30–40	
Motorcycle	Gasoline moped/ scooter	3–5 (50–125 cc)	50–80	2–3 L	120–200	
	Gasoline motorcycle	4–6 (100–125 cc)	60–80	2–3 L	120–200	

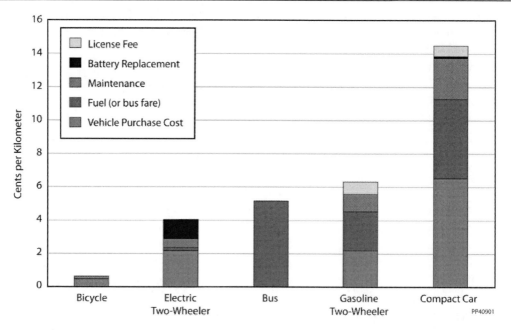

Figure 1-1. Cost of Common Transport Modes in China.

1.4. Electric Two-Wheeler Issues

The amount of solid waste that results from operating ETWs is significantly higher than the amount from operating motorcycles. A life-cycle emissions study comparing an ETW to a motorbike concluded that an ETW generates 2.7 g/km of solid waste (63% from coal combustion and 14% from battery disposal), while a motorbike generates only 1 g/km. It is estimated that emissions of lead from using an ETW are 0.05–0.10 g/km; they result from inefficiencies in the dispersed, small-scale lead production and recycling process.

Traffic safety is perhaps the most important issue with regard to ETW growth. In November 2006, Guangzhou became the third city in China (after Fuzhou and Zhuhai) to ban ETWs, acting on advice from the traffic management bureau, which cited traffic safety concerns. Conversations with traffic police indicated that the concern over safety is mainly a result of the erratic driving behavior of two-wheeler drivers, which affects other vehicle drivers and traffic efficiency. The high speeds, low weight, and relatively silent operation of ETWs also pose a threat to bicyclists riding in the nonmotorized-vehicle lane. Thus, automobile owners and bicyclists often perceive ETWs negatively. The safety issue of ETWs when mixed in traffic has been a key consideration during the drafting of new Chinese national ETW standards, which have been under intense debate during revision. Electric bikes are also not the most efficient users of scarce road space. While ETWs can move more people per lane than cars can, they move fewer people per lane than buses can. In Taiwan, ETWs were promoted between 1996 and 2003 as a means of improving urban air quality, but that effort failed. The main problem was that the scooters were too expensive because of their high power and energy requirements.

Table 1-2. Pros and Cons of Using ETWs.

Pros	Cons
Zero tailpipe emissions	Use electricity, 75% of which is produced from coal (i.e., dirty) in China
Energy efficient (70–80 km/kWh)	Emit lead during battery production and recycling
Inexpensive; save human energy that can then do other work	Could result in less efficient flow of traffic than does use of public transit
Convenient; can be "refueled" at home or work	Elicit safety concerns when mixed together with other vehicles (quiet, fast, heavy, poor brakes)

1.5. Gasoline Scooters and Motorcycles

Motorcycles in China come in three main styles: scooter, underbone, and traditional motorcycle (or horseback-type); there are also a very few mopeds. The following classification is helpful in characterizing the wide range in motorcycle types:

- *Mopeds.* These are small, light, inexpensive, and efficient for riding around town and are usually started by pedaling (motorcycle + pedals = moped).
- *Scooters.* These are motorcycles with a step-through frame and generally smaller wheels than those of a traditional motorcycle. They can be ridden without straddling any part of the bike. Scooter-style motorcycles are usually equipped with automatic transmissions
- *Underbones.* These are small motorcycles, between a scooter and a true motorcycle, with a step-through frame. The fuel tank is under the seat. They are popular in Southeast Asia. Underbones are also known as "cub"-style motorcycles, since they are based on the original cub motorcycle introduced by Honda in the 1960s.
- *Standard motorcycles.* These have tear-shaped fuel tanks at the top and just behind the instrument panel.

LPG scooters are popular in Shanghai because they are exempt from the city-side motorcycle ban. However, this type of scooter is excluded from the analysis, since it is exclusive to Shanghai.

Motorcycle engine types and styles have changed since the early 1990s. During the first half of 1990s, the most prevalent motorcycles were two stroke, 110 cubic centimeters (cc) or less, and horseback type (standard). In the latter half of the 1 990s, the market share of four-stroke scooters with engines that were 125 cc or more in size increased sharply. From 2000 onward, the popularity of underbone frame types increased, and they became the most common motorcycle engines in Southeast Asia (especially the ones based on the Honda C 100). By 2002, as a result of tightening environmental regulations, only a few models of two-stroke motorcycles were available. The market share of engines (by engine displacement size) in China in 2002 was as follows: 125 cc accounted for ~45%, 110 cc or less accounted for ~28%, and 50 cc or less accounted for <8%. The market share by type was as follows: four-stroke standard motorcycle was ~37%; four-stroke scooter style was ~30%; and underbone (18%), two-stroke motorcycle or scooter style was ~1 1%. By 2002, motorcycle engines had converged to three dominant models — the C 100, CG125, and GY6 — all of which are four-stroke engines.

The four largest national markets for motorcycles in 2006 in order of annual sales volume were China (14.6 million), India (8.2 million), Indonesia (4.6 million), and Vietnam (2.3 million). In terms of vehicle ownership in 2002, there was one motorcycle for every 2 people in Taiwan, for every 4 people in Thailand, for every 7 people in Vietnam, for every 15 people in Indonesia, for every 16 people in China, and for every 63 people in Malaysia. Since 1998, motorcycle ownership grew faster in rural households than in urban households, with 32 owners per 100 rural households compared with 24 owners per 100 urban households in 2003. This trend was likely due to the motorcycle bans in many cities, which also started in the late 1990s. In 2006, China produced half of the world's motorcycles. Most exports from China are sold to the low-end market (Southeast Asia and Africa).

1.6. Electric Two-Wheelers around the World

A huge majority of the worldwide ETW market — 95.8% — is concentrated in China. The ETW market in Japan, Europe, and more recently, India, is small but growing. The market is 1.4% in Japan, 1.5% in Europe, 0.3% in India. The market is only 0.8% in the United States and 0.2% in Southeast Asia.

After China, the next largest ETW market is Japan, with annual sales of 270,000 bikes/yr in 2006 and 13% average annual growth since 2000. Pedelecs (a style of ETW driven primarily by human power with battery assist) are the dominant type of ETW. Most pedelec ETWs use Ni-MH or Li-ion batteries. Battery capacity ranges from 0.2 to 0.6 kWh, motor sizes range from 150 to 250 W, and prices range from $700 to $2,000. In Europe, the market was estimated at 190,000 bikes/yr in 2006. Electric bikes in Europe are also mainly pedelec style. Sales in the Netherlands are the greatest because of its extensive bicycle infrastructure and deep-rooted biking culture. Germany and Belgium are the next largest markets for pedelecs.

India's electric bike market is small, but forecasts for growth are optimistic. In other developing countries throughout Southeast Asia, like Thailand, Vietnam, and Indonesia, where two-wheelers are the dominant form of transportation, ETWs have not gained a significant market share. This trend may be attributed to the fact that valve-regulated lead-acid (VRLA) battery performance (i.e., range and lifetime) degrades quickly in areas where temperatures are very high (or very low) throughout the year. Gasoline-powered motorbikes are the dominant mode in the larger cities of these countries. In the United States, the very limited electric bike market consists mainly of recreational riders who rely on the assistance of the electric motor out of physical necessity. The ETW is not a common commuter vehicle in most cities because commuting distances are long, the bicycle infrastructure is virtually nonexistent, and most bicycle commuters do so primarily for recreation.

1.7. China's Changing Transportation

The development of bicycles as a transportation mode in the past 50 years in China can be divided into three distinctive eras. The first era, from the early 1 960s to the early 1 980s, was marked by the steady growth of bicycles as the supply struggled to meet the demand. The second era, from the early 1980s to the 1990s, covered the entire cycle of rapid growth, saturation, and decreasing production. The third era, since the 1990s, has reflected a new cycle of bicycle quality, style, and replacement market. During the first era, the country was administrated by the centrally planned economic system, and the bicycle industry developed slowly, especially with respect to the iron and steel, machine, and electronics industries. Mixed land use provided a fertile ground for bicycle traffic to flourish. Most people who worked for the government or government-owned factories lived in unit-owned dormitories, which were usually built near the working sites. Most of the travel distances were within walking range. Most of the reasons for trips were to go to and from work and school. During this period, bicycles were the dominant mode for passenger transport, even though it was not easy for an ordinary family to buy a new bicycle at that time. The low level of income, dense and mixed land-use patterns, and the short distances to be traveled were the main reasons for the formation of nonmotorized traffic in China.

The second era of bicycle traffic development in China was marked by the beginning of "economic reform." During this transition period, changes took place in many sectors of the economy, as experiments with bold, new policies, which were similar to those of a market economy and deviated from the former centrally planned economy, took place. In 1987, a milestone was reached when 40 million bicycles were produced in China. Subsequently, most families in urban areas had the ability to purchase new bicycles as a result of the obvious increases in income in China. The price of a bicycle changed from 150 ¥ in 1973 to about 500 ¥ in the 1 990s. The bicycle ownership rate kept pace with the growth in income during this period. In addition, the Chinese central government eliminated the bicycle tax in 1977, and many local governments and companies provided subsidies for bicycle owners. These economic policies boosted the development of bicycles in Chinese cities.

It is worth noting that almost all cities in China started to use the "three sections" cross-sectional design for their streets. There is a separate section of the road for bicycles that is between the innermost lanes, which are for automobiles, and the outermost lanes, which are for pedestrians. This three-section urban street design helped increase bicycle traffic in Chinese cities.

The third era was marked by a fluctuating bicycle share and heated debate on the fate of bicycle traffic. A decade of economic reform and positive results swept away any doubt of the viability of such reform. After more than 10 years of high production, bicycles saturated most cities, and production started to decrease in 1996. The level of bicycle production in 1999 was similar to that of the early 1 980s. Only 1 million bicycles were sold in that year. Bicycle ownership remained stable, even though income continued to grow.

In this era, urban populations and areas expanded quickly, driven by the rapid economic development. Urban structures and land-use patterns changed considerably. With the development of society in China, travel distances became much longer than before. The reasons for trips increased, and trip times increased also to a certain extent. The number of private automobiles increased rapidly in large cities, especially in the last five years. In China, motorized traffic has become an inexorable trend, and the proportion of motor vehicles in the various transport modes increased steadily.

During this time, public transit has developed slowly and automobile traffic is still in its infancy, even with its explosive growth rate, so there is still room for bicycles to play a dominant role in urban passenger transport. Nevertheless, the mode share of bicycle traffic has shifted up and down, depending on policies that have been implemented during different administrations and in different regions. The bicycle share has fluctuated along with the shifts in urban transportation policies, service levels provided by transit, and other market conditions. For instance, Nanjing and Shijiazhuang, two large cities that are both capitals of their respective provinces, experienced dramatic shifts during the past two decades.

Despite the introduction of private automobiles and mass rail transit in a few selected cities, the bicycle still maintains its unique and dominant position in urban transportation. In 2004, the number of bicycles in Beijing had reached 10 million. Similar situations exist in most Chinese cities; there were about 450 million bicycles in China in 2004, which translates to 40 bicycles per 100 people. The rate of bicycle ownership was higher in large cities than in medium or small cities.

1.8. Background on Electric Two-Wheelers in China

The basic principle of electric bicycles is that the electric energy is supplied by batteries and that the motor drives the bicycle. Electric bicycles can be divided into two types. The first type is solely electric; the rider does not need to exert power. Most of the electric bicycles that are produced in China are this type. The second type involves some peddling (friction drive); electricity is supplied from the electric motor on demand to reduce the foot pedal resistance. It is being developed by Tsinghua University. Both types of electric bikes are produced by Mt. Kun Company, a subsidiary of the large company, Taiwan Province.

The number of electric bikes in China is small. The profile of an electric bike is similar to that of an ordinary bike. In June 1997, a general safety standard for electric bikes, QG2302.97, was issued; it began to be implemented in April 1998. This standard stipulates that when the speed of an electric bike reaches 18 km/h, the bike's dry braking distance should be no shorter than 4 m, and its wet braking distance should be shorter than 15 m. It also stipulates that the frame/front fork assembly's shock strength load capacity should be 70 kg. The electric bike should go no faster than 20 km/h.

Electric bicycles have five major assembly components: electrical machinery, a controller, batteries, a battery charger, and a frame. The batteries that can be used in E-bikes are mainly lead-acid, Cd-Ni, iron-nickel, hydrogen-nickel, and Li ion batteries. In the current market, 65% of E-bikes use lead-acid batteries, 30% use Cd-Ni, and 5% use hydrogen-nickel or Li-ion batteries. Table 1-3 gives basic characteristics of electric bikes.

1.8.1. Types of Electric Bikes

There are three types of electric bicycles in the Chinese marketplace, as follows.

- *Deluxe electric bicycles.* These motorcycle models have an attractive exterior and are comfortable to ride, luxurious looking, and expensive. They are sold in large quantities. However, the technology is not very mature at present. Their bodies are low and not suited for driving on curvy and uneven roads.
- *Simple electric bicycles.* These have structures similar to those of pedal bicycles; they are very simple and agile. European and American countries are focusing on pursuing these popular bikes. Their construction is simple, yet they are comfortable.

Their design is also simple. These are the main types of bikes being sold in China's northern sector.
- *Neutral electric bicycles.* These are between the other two types. They are stylish and designed with luxury in mind. They are not too short, comfortable to ride, and give riders a good view of the road. Chinese consumers with the means to purchase them really like them. Chinese experts suggest that this type of E-bike will lead the worldwide market in the future.

To meet consumer needs, innovation is required in the development of electric bicycle products. Chinese experts believe that innovations similar to those found in digital and nanotechnologies must come to electric bikes if they are to become popular in China.

1.8.2. Advantages of Electric Bikes

Electric bicycles are considered new transportation vehicles in China. They have these advantages:

- They have zero emissions and do not pollute the atmosphere.
- They cost only about 2,000–3,000 ¥ and consume only about 1 kWh of electricity to travel 100 km; they are affordable to the Chinese consumer.
- They can go about 30–50 km on a single charge. This capability meets the needs of an average, urban, bicycle-riding office worker, which should result in a large market demand.
- When their speed is kept at 15 km/h (which is especially easy for an intelligent E-bike, such as one with an electronic speed controller that automatically cuts off the speed when it goes above a set point), they do not pose an excessive threat or safety problem to riders of manually operated bicycles or pedestrians.
- Most of their spare parts are the same as those used in manually operated bicycles; they are easy to repair.
- Most are used during the day and can be charged late at night, which is advantageous in balancing the electric power load.
- With massive marketing and their large export potential, electric bicycles could lead to a new industry in China and make a valuable contribution to the country's economic growth.
- Both all-electric and on-demand electric bikes reduce human exertion. They are structurally simple, convenient to charge, and easy to operate. Users do not need a driver's license to operate them, and the bicycles do not require an annual inspection.
- They are compatible with both city and country roads; roads can be made of cement, asphalt, or mud.
- They improve the quality of people's lives and are convenient for travel.
- They will reduce traffic congestion in large cities.

Table 1-3. Basic Characteristics of Electric Bicycles.

Characteristic	Ordinary Bicycle	Electric Bicycle
Vehicle speed (km)	10 to ~15	15 to ~20
Sphere of action (km)	15	25 to ~40
Should operator reduce effort?	No	Yes
Average price (¥)	300	1,500

Source: Chinese Bicycle Association

- They will help reduce GHG emissions and pollution in large and mid-size cities worldwide.
- They will help reduce energy consumption. Their annual operating cost is about 15% of that of motorcycles and 20% of that of mid-sized vehicle running on gasoline. The energy consumed by a motorbike is about 10% of that consumed by a mid-sized gas-fueled car.

1.9. Electric Two-Wheelers in China

E-bicycles have become popular in many Chinese cities because they provide benefits to riders in terms of private mobility, accessibility, time savings, and low cost and because they are environmentally friendly. From the early 1 990s, when E-bicycles emerged in Chinese cities, there were disputes about their use. Because E-bicycles have a power system, some policymakers asserted they should belong to the category of motorized vehicles; others rejected this classification. In 2003, the National People's Congress of China passed the Law of the People's Republic of China on Road Traffic Safety (LPRCRTS), which became effective on May 1, 2004. It classifies electric bicycles as nonmotorized vehicles from an operational and regulatory perspective. The law's technical standards stipulate that E-bicycles cannot travel faster than 15 km/h or weigh more than 40 kg. The law does not require an operator to have a driver's license or helmet. These stipulations mean that E-bicycles can share the slow-speed, no- motorized lane with bicycles and cannot be operated on motor-vehicle lanes. However, in consideration of the variety of safety issues and impacts on urban transport systems associated with E-bicycles, the law grants provincial and municipal authorities the right to determine whether E-bicycles can (or cannot) be operated on local streets. Different cities have different policies on the development of E-bicycles. Some big cities, such as Guangzhou and Fuzhou, forbid their use, while other cities, such as Chengdu and Kunming, promote them. Most cities adopt a neutral policy; they neither promote nor restrict E-bicycles.

Even though there is no clear definition or stable policy in many cities, the use of E-bicycles has grown rapidly in China. More than 1,000 E-bicycle makers produce thousands of models. In 2006 alone, China produced 19.5 million E-bicycles. E-bicycles in China can be categorized as bicycle-style electric bikes (BSEBs) or scooter-style electric bikes (SSEBs). BSEBs provide a functional pedal and typically have 36-V batteries and 1 80—250-W motors. SSEBs usually have 48-V batteries and 350–500-W motors and look more like motorcycles than bicycles. According to the Chinese National E-Bicycle Standards (1999) — which are standards for product quality — regulations prohibit E-bicycles from going faster than 20 km/h; however, most E-bicycles, especially the scooter-style ones, can go at a faster speed. To meet consumer demands for faster, more comfortable E-bicycles with a longer range, some manufacturers increased the power, maximum operating speeds, maximum weight, and other parameters of E-bicycles. One common method manufacturers use to do this has been to provide the E-bicycles with speed-limiting devices that can be easily removed after purchase.

The fast development of E-bicycles challenges the adaptability of urban transport systems. Even though some local traffic management authorities refuse to release licenses to scooter-style E-bicycles, these SSEBs are still sold and enter the urban transport system. Their presence enlarges the dynamic performance differential between electric bicycles and general bicycles, as well as the operating speed differential.

In 1998 (the first year there were national statistics), the Chinese annual output was 58,600 bikes; this number had jumped to 21,300,000 bikes in 2007, which accounted for 43% of the Chinese economy's average annual growth during that year.

In 2003, when China's national electric bicycle output was 3,997,200, output from Tianjin, Jiangsu, and Zhejiang Provinces was 2,809,700, which accounted for 70% of ETW production in China. In 2007, the output from the above three provinces was 15,845,000 ETWs, which accounted for 74% of total ETW production in China. In Tianjin in 2007, the annual output was 6,461,000, which accounted for 30% of total national bike production.

There are 10,000 large and small companies involved in the national production of electric bikes. The small and mid-sized companies accounted for 35% of total national bike production during 2007. Most of these bikes used lead-acid batteries. During the same year, the entire industry's production of E-bicycles running on Li-ion batteries surpassed 100,000 ETWs. In 2007, exports of electric bicycles were about 395,000; exports from Japan, the United States, and the EU were 203,300, which accounted for 58% of their production. Japan, the United States, Italy, Holland, Germany, Hungary, and Great Britain imported E-bicycles from China that accounted for 87,800 ETWs.

The year 2007 was frustrating for those involved in the E-bike industry in China. The cost of raw materials rose by a factor of three and the cost of labor increased, making it very expensive to manufacture bikes. Numerous E-bike manufacturers and dealers cut back their production. The industry needs to implement innovations and cut manufacturing costs to stay profitable. In Tianjin, a new area is opening up for developing and manufacturing bicycles, which has brought about an upsurge in investment. As the Tianjin E-bike industry catches up to the rest of the Chinese bike industry and establishes an industrial base, production will grow and more ETWs will be available for domestic use and export purposes.

1.9.1. Chinese National Support for Electric Bikes

The E-bike market emerged in the late 1 990s and has grown considerably up to the present. The following chronological list describes important events in the history of E-bikes in China:

- In 1987, the Electric Vehicle Institute of China Electro-technical Society was founded.
- In December 1991, the Chinese National Science Board named electric bike development as one of 10 technology projects. The electric bike was still deemed a major technology project in the period around 1995.

- In 1994, China's National Defense Science and Technology Committee (NDSTC) and the U.S. Department of Defense (DOD) signed a military privatization collaboration agreement. One item in this agreement was to work together to promote and apply EVs.
- In 1995, Prime Minister Li Pong and Vice Prime Ministers Li Lan-Qing, Wu Bang-Guo, and Tai Jia-Hua successively made important comments on the development of the EV industry. On the basis of these comments, in
- October 1995, the Chinese Mechanical Industry Department held a seminar to create an EV development strategy together with the National Science Board and Economy and Trade Committee.
- In April 1999, China held a meeting entitled "Clean Air Program — Clean Automobile Action" in Beijing. Vice Prime Minister Wu Bang-Guo attended the meeting and made an important speech. Thereafter, China founded the Coordination and Leading Team for Clean Automobile Action.
- In October 1999, the 16th International Electric Vehicle Meeting took place in Beijing. Both Vice Prime Minister Wu Bang-Guo and Deputy Director of the Technology Department Xu Guan-Hua spoke at the meeting.
- In 2000, administrative coordinators of NDSTC and DOD recommended that the U.S. Department of State actively promote the use of E-bikes in China and the United States.
- In July 2000, the Legislative Office in the U.S. Department of State and the Traffic Control Bureau in China heard comprehensive reports by the China Bike Association on the status of E-bikes in China. They agreed that the speed limit would be 20 km/h and that E-bikes with a pedaling function could be treated as nonmotorized vehicles.

1.9.2. Shanghai and Electric Bikes

In Shanghai, cars and trucks are crammed on the elevated highway that cuts through downtown. On the smaller roads below, traffic moves at a steady 10–15 km/h. It includes an assortment of two- and three-wheeled vehicles — everything from simple steel-frame bikes and heavily laden pedal-powered carts to motorized scooters and electric bikes. There are an estimated 1 million electric two-wheelers on Shanghai's streets. Despite China's growing infatuation with automobiles, people in the world's most populous nation continue to move primarily on two wheels, and, increasingly, an electric motor drives them. The China Bicycle Association, a government-chartered industry group in Beijing, estimated that during 2007, manufacturers sold 9.5 million E-bikes nationwide — nearly double the sales in 2006 — and would probably ship more than 14 million E-bikes in 2008.

There has been a huge desire for motorized personal transportation in China as its cities have sprawled. E-bikes are an attractive option for commuters, service people, and couriers. At 1,500–3,000 ¥ (U.S. $180–360), an electric bike can be bought at a small fraction of the cost of an automobile. Riding an E-bike can be exhilarating. An electric motor built into the hub can propel you to speeds of 20 km/h or more. However, despite the appeal of electric bikes, some Chinese cities have banned them, alleging environmental drawbacks and concerns about public safety. But that has not stopped millions of people in China from buying ETWs. Such development is astonishing to ETW advocates in the United States and Europe, who have struggled for a decade to build a market for E-bikes.

A blend of necessity and opportunity kick-started China's first E-bike manufacturer, Shanghai Crane Electric Vehicle Co., based in the Pudong section of Shanghai. The company descended from a venture-capital arm of the Shanghai government that had been investing in electric-drive technology in a bid to lead a new national electric-automobile R&D program. When Shanghai lost the automobile research bid to Guangzhou in 1994, Shanghai's EV team turned to electric bikes.

Testing of 100 prototype bikes made by Shanghai Crane in 1995 revealed that they needed a lot more development. In barely three months of use, the motors burned out and the lead acid batteries — designed to be removed from the bikes and taken inside for plug-in charges — no longer could take a charge. However, the testers thought the bikes were fun to ride and handy for carrying parcels, which suggested that a more durable product would find a ready market. When Shanghai banned sales of gas scooters (and their polluting two-stroke engines) in 1996, Crane spun out from R&D to fill the market void. In 1997 the company's first products were rolled out; they were conventional bike frames outfitted with a 150- or 1 80-W hub motor in the front wheel; a 24-V, 7-A•h (ampere-hour) lead acid battery on the rear rack; and a simple electronic controller on the handlebars.

Their performance was much better than that of previous bikes: The motors went well beyond the three-month mark, and the batteries, now rated for about 300 charges, could carry the bike as far as 50 km on a charge with minimal pollution. E-bikes can carry a single driver with 15–20 times greater efficiency than that of an average small car. As a result, they generate just a fraction of the air pollution and carbon dioxide (CO_2) emitted by cars. Sales mounted, and the success of Crane attracted competition, bringing both start-ups and conventional bike manufacturers, such as T and D Continental Dove Company of Nanjing and Shanghai Forever Co., into the market. Today, the China Bicycle Association estimates there are an astounding 1,500 companies manufacturing electric bikes, many of them local operations producing a few thousand bikes per year.

Producing 50,000 bikes a year with a workforce of 210, Crane is one of the few businesses that can sustain an R&D operation. However, because of China's weak protection of intellectual property, the innovations made by companies like Crane spread quickly to the entire industry. Cranes believes that more R&D is needed to improve products. Better bike technologies include brushless motors that deliver higher torque, electronic controllers, and lead- acid batteries that deliver a range of up to 60 km and last up to 2 years.

The look of the ETWs has changed. They are more stylish; have large platforms for resting feet as well as packages; have few pedals (or none at all); and, in some cases, have more powerful batteries and motors that boost the top speed to close to 30 km/h. These electric scooters accounted for roughly two-thirds of the 10,000 EVs Crane sold last year. The owners are generally commuters whose trips have lengthened as the city has grown during the last decade, delivery and salespeople who crisscross neighborhoods, elderly men and women running low on pedal power, expectant mothers, and students. They all want a faster, easier ride than they get with a conventional bike.

Automotive and motorcycle manufacturers, transit operators, and some government officials, however, have slowed or stopped the growth of the electric bike in major cities such as Beijing and Guangzhou. Even the China Bicycle Association, which purportedly represents bike makers, has sought to discourage manufacturers from adopting faster scooter designs.

Despite the electric-bike industry's decade-long history and commercial success, it was only last year that China's National People's Congress amended the national road safety law to officially give riders of electric bikes a right to use the roads. The legislation legally equated the electric bikes with conventional bicycles — wherever bikes can go, electric bikes can go. But the amendments include an important caveat: Municipalities have the final say on whether to allow electric bikes in their localities. Some have refused to do so. In rejecting electric bikes, the municipalities cited such concerns as the threat of pollution from spent lead-acid batteries; interference with automobiles, resulting in accidents or slower traffic; and the impact on the viability of public transit systems. Advocates for green transportation say these arguments amount to thinly veiled attempts to protect the electric bicycle industry's competitors. None of the arguments against electric bikes have merit. Lead-acid batteries are also used in cars, and the real pollution source is not the electric bikes but the automobiles. Transit operators and manufacturers should be forced to compete with electric bikes so they would offer more efficient services and cheaper, cleaner vehicles. The problem is that electric bike manufacturers are insignificant when compared with the other interest groups, particularly car makers, who are attracting billions of dollars of foreign investment. The automotive industry is identified as a "pillar industry" in China's official five-year plans.

Although the odds against them are daunting, E-bike manufacturers are pushing forward with surprising success. Like Crane, Luyuan EV was a government venture-capital spinoff. Nine years ago, with the help of its initial investors, it developed a prototype bike. In 2007, Luyuan EV sold 120,000 electric bikes and scooters, and it expected to sell 300,000 ETWs in 2008. Luyuan EV is located in Jinhua, an industrial metropolis with a population of 1.4 million. It is south of Shanghai and located in Zhejiang Province.

Conflict over electric bikes is not limited to the municipalities and the manufacturers. Even the China Bicycle Association has been clashing with some companies, including Luyuan, over what types of ETWs should be on the road. The bike group enforces a national standard for electric bicycles, and no matter what parameter you choose (maximum weight of 40 kg, width of 220 mm for the pedal shaft, maximum speed of 20 km/h), many of the latest electric scooters cannot match these parameters or standards.

Many electric scooters, for example, are outfitted with nonfunctioning pedals and with speed-limiting devices designed for easy removal after purchase. Luyuan's latest machine just meets the E-bike standard. Luyuan calls its new product the LEV, short for light electric vehicle. The LEV weighs 95 kg; its 48-V, 20-A·h battery has two times the energy of a standard bike; and its 500-W motor controlled by a central processing unit (CPU) propels it to 35 km/h.

The China Bicycle Association is concerned that vehicles that violate the standard could damage the electric bike industry. It fears a regulatory backlash could result if riders of powerful ETWs like LEVs were seriously injured in accidents. Such a reaction would hurt the entire industry by undermining the justification for allowing electric bicycles on the road. If the electric bicycle were to be more like the motorcycle, the ability to classify an electric bike as a bicycle could be lost. The China Bicycle Association is pushing for amendments to the national electric bike standard to close its loopholes. But Luyuan and other manufacturers have other ideas; they advocate revisions that would boost the electric bike's top speed to reflect current consumer demand. At the moment, the debate is gridlocked, and vehicles such as the LEV keep rolling off assembly lines and onto China's busy, crowded streets.

The biggest challenge facing electric bike makers may not be municipal bans, conservative standards, or even technology. It may be the roads themselves. China's development is following the path of Western countries; the country is rapidly redesigning its cities around the automobile. Across China, cities are rejecting a mixed-use model and redeveloping along a strict zoning model. China is razing residential buildings in center cities to make way for office towers and is paving farmland on the periphery to create large industrial parks. Displaced from the urban centers, houses and other residential buildings are springing up in sprawling suburbs. The automobile is king in this model, because in the absence of extensive public transit, cars are the only way to get from distant suburbs to offices and industry parks.

To make way for more cars, China's cities are widening their main roads and building highways. The result has been a rapid increase in the number of automobiles, which, just as they do everywhere else in the world, almost instantly absorb the extra roadways. The resulting gridlock has been especially acute in China's capital. Beijing had 1 million cars in 1997 and 3 million in 2008. An urban planning expert at Beijing University believes that wider roads are more efficient for traffic. However, if the electric bike market were to expand, the wider roads might not be needed. A "car culture" is a disaster for bicycles. Road widening often comes at the expense of bike lanes. Highways are off-limits to bikes and nearly impossible to cross. On smaller roads, rush-hour traffic blocks bike lanes and intersections, prompting outbursts of road rage from frustrated cyclists.

China's oil imports are on the same exponential growth path as are its car fleet. China has eclipsed Japan as the second-biggest importer of oil, bringing it into direct competition with the world's leading consumer of petroleum: the United States. To lessen import dependence and environmental burdens, China has promulgated fuel-efficiency standards that are similar to the European standards. It is considering imposing a 20–50% national tax on retail gasoline and diesel fuel.

1.10. Gasoline Two-Wheelers in China

1.10.1. Number in Use

According to the PRC's National Environmental Protection Administration's *Report on the Environment for 2005*, ownership of automobiles and motorcycles exceeded 43 million and ownership of motorcycles exceeded 94 million by the end of 2005. Compared to 2004 figures, the number of automobiles increased by 20.6%, while the number of motorcycles increased by 23.6%. Private car ownership showed high annual rates of increase (23%), bringing the number of private vehicles in the PRC to 14.8 million, or about 55% of the total number of vehicles in the previous year, 2004. The high growth rates were directly correlated with the growing economic prosperity in Chinese cities. In Beijing, the vehicle fleet quadrupled from 0.5 million in 1990 to 2 million in 2002. This rapid rate of motorization is expected to continue in the next decades. It is generally expected that by 2020, the total number of four-wheeled motor vehicles in the PRC will be between 100 million and 130 million. The total number of motor vehicles in the PRC could reach 248 million by 2015, with the highest rate of increase occurring for cars and sport/utility vehicles (SUVs), followed by two-wheelers. The number of two-wheelers is expected to decline after 2025, when personal incomes will have reached a level that will allow people to purchase a car instead of a motorcycle.

The Chinese urban transportation system is dominated by private cars, buses, taxis, motorcycles, scooters, and bicycles. Most cities, such as Beijing, Guangzhou, and Shanghai, still have dedicated and separate lanes for bicycle traffic in urban areas. These lanes, however, are not sufficiently integrated into the whole transportation network. An investigation of trip patterns in Chinese cities indicates that the largest share of trips is still made by walking and cycling (65%), followed by public transport (19%) and private motor vehicles (16%). However, the number of walking and cycling trips is expected to decline as the number of trips by private motor vehicles begins to increase. In Xi'an, motorcycles number about 14 million but take up only 5% of the total share of modal trips in the city. Walking, bus trips, and bicycle trips play a major role in modal trips, with shares of about 22% for walking and 33% for bus trips and cycling.

1.10.2. Emissions from Mobil Sources

Estimates of emissions per vehicle type in 2005 are presented in Table 1-4. Particulate matter (PM) emissions are mainly from ETWs followed by heavy freight trucks, while NO_x emissions are mainly from heavy freight trucks followed by minibuses, paratransit (small) vehicles, and buses. The emissions might be a result of the older engine technology used in these types of vehicles. The rapid increase in the number of four-wheeled vehicles will lead to increases in the contribution from four-wheeled vehicles and a relative decrease in the contribution from two-wheelers.

The PRC adopted a road map for new vehicle standards and laid out a schedule to introduce vehicle emission standards equivalent to the EU emission standards for light-duty vehicles. In December 2005, the State Council of the PRC approved the implementation of Euro III in 2005 and of Euro IV in 2007 for light-duty and heavy-duty vehicles in Beijing. The State Council required Beijing to ascertain the availability of fuel of corresponding quality by the time of implementation.

The production and sales of electric bicycles and scooters soared rapidly in the 2003–2008 period. Annual electric bike sales in the PRC grew from 40,000 in 1998 to 10 million in 2005. The increase was largely brought about by items of legislation that banned gasoline-fueled scooters and bicycles that were introduced beginning in 1996 in several major Chinese cities, including Beijing and Shanghai. Electric bikes are gaining an increasing share of two-wheeled transportation in the PRC. In Shanghai, there are an estimated 1 million ETWs. In cities such as Chengdu and Suzhou, the share of electric bikes has reportedly surpassed the share of regular bicycles. Electric bikes, which use lead batteries as the main source of stored energy, are touted as a zero-emissions form of transportation that can help improve urban air quality. However, the environmental impacts from lead emissions (the lifetimes of the lead batteries used for E-bikes are limited and the batteries must be disposed of) may negate some of the benefits derived from the absence of tailpipe emissions. A comparison of electric bikes and other power- assisted vehicles is given in Table 1-5.

1.11. Energy Consumption in China

Measurements have shown that the average electric bicycle uses 1.5 kWh of electricity per 100 km. When 10,000 km are traveled per year, the total cost of electricity cost is about 90 ¥, which is 15 times less than the cost of running a motorcycle and 40 times less than the cost of running car, in terms of energy consumption cost per unit of travel distance, as shown in Table 1-6.

Table 1-4. Total Pollutant Emissions from on-Road Vehicles by Vehicle Type in 2005.

Vehicle Type	PM Emissions (1,000 metric tons)	NO$_x$ Emissions (1,000 metric tons)
Light-duty vehicle	37.1	175.4
Medium freight truck	22.9	285.8
Heavy freight truck	36.2	483.2
Two-wheeler	61.9	120.7
Three-wheeler	0.0	0.0
Bus	19.6	260.8
Minibus and paratransit	24.9	330.8
Total	202.7	1,656.6

Table 1-5. Comparison of Electric Bikes and Other Power-Assisted Vehicles.

Parameter	Fuel-Assisted Vehicle	Gas-Assisted Vehicle	Electric Vehicle	Electric Bike
Vehicle weight (kg)	70–80	70–80	90–100	35–40
Maximum vehicle speed limit (km/h)	24	24	24	20
Continuous driving distance (km)	150	150	50–60	50–60
Charging (oil-filling) time	Several min	Several min	5 to 8 h	5 to 8 h
Type of energy used	Petroleum	LPG	Battery	Battery and human power
Driving noise (dB)	65–70	65–70	55–60	55–60
Unit price (10^3 ¥)	4–10	4–6	5–8	2–3
Driver's license test required	Yes	Yes	Yes	No
Helmet required	Yes	Yes	Yes	No
Age limit	Yes, over 18	Yes, over 18	Yes, over 18	No

Table 1-6. Comparison of Energy Consumption per 100 Kilometers by E-Bicycles, Motorcycles, and Cars.

Parameter	E-Bicycle	Motorcycle	Car
Energy source	Electricity	Gasoline	Gasoline
Amount of fuel consumed	1.5 kWh	3 L	10 L
Cost (¥)	0.8	12	40

1.12. Electric Two-Wheeler Market and Production in China

The PRC is the world's biggest producer of bicycles. In 2004, 79 million units were manufactured, of which 22–25 million were for the Chinese domestic market. This number was down from a peak of 40 million just a few years before. The Chinese Government policies that promoted cars had a discouraging effect on cycling, and the removal of many bike lanes was thought to be necessary to accommodate an increasing use of cars. In Beijing, only 20% of the commuters rode bikes in 2002, compared to 60% in 1998. However, many cycle paths were built in the "old town" areas and newly built communities for the 2008 Beijing Games. While bicycle sales were waning, electric bicycles increased in popularity. Some 260 companies in the PRC are estimated to be making electric bikes and their components. About 7.5 million electric bicycles were sold in the PRC in 2004; the number rose to about 9 million in 2005, even though they were banned in some key cities. The output of electric bikes is expected to maintain an annual growth rate of at least 80% in the years until 2010. These bikes provide an attractive option for commuters, service people, and couriers who have a need for motorized personal transportation. Their dramatic growth has been largely a result of legislation banning gasoline-fueled scooters and bicycles that began in 1996 in several major Chinese cities, including Beijing and Shanghai. The Government concluded that electric bicycles were not in keeping with Beijing's image as a major world capital and recommended that they be phased out. The Beijing Municipal Public Security Bureau announced in 2002 that electric bicycles would be forbidden Electric bicycles come in many versions, and there is fierce brand competition (Appendix C). Market leaders, such as Fushida, Yadi, and Xinri, hold a share of about 70%. The E-bikes have a top speed of 20–30 km/h and a range of 25–1 00 km. During operation, they emit zero local air pollutants, but they do use about 2 kWh of electricity per 100 km. Their power range is 200–600 W, and they take about 6–8 hours to charge.

According to the China's State Statistical Bureau, the country's bicycle production in 2007 was 65,497,775; it had grown at a rate of 5.12% over the same period in 2006. The country's regional output of bikes in 2007 is shown in Table 1-7.

In the first quarter of 2008, bicycle imports grew by 123 .08%, in comparison with the same period the previous year.

According to the National Customs Administration, in the first quarter of 2008, China's bicycle imports grew by 3%, compared to 31% during the first quarter of 2007. Total bike imports amounted to U.S. $195,000, which represents growth of 254.60% over year 2006 imports. Also in the first quarter of 2008, imports of spare parts jumped by 50 tons.

According to China's Automobile Industry Association, in the first quarter of 2008, the number of motorcycles exported by the top 10 Chinese companies was as shown in Table 1-8. They exported 1,245,500 motorcycles, which accounted for motorcycle exports of 57% of the previous year's exports. Also according to China's Automobile Industry Association, in April 2008, 2,540,500 motorcycles were produced and 2,524,700 were sold; the result was a growth rate of 8.58%. The previous month's rate was 6.06%.

Table 1-7. Number of Bikes Produced in Regions of China in 2007.

Region	No. of Bikes Produced
Tianjin	17,110,952
Shanghai	6,401,915
Jiangsu	10,776,395
Zhejiang	15,115,135
Anhui	2718
Jiangxi	2,682
Shandong	589,321
Henan	689,155
Hubei	86,775
Hunan	1109
Guangdong	14,189,096
Guangxi	73,349
Sichuan	438,805
Shanxi	20,368

Table 1-8. Motorcycle Exports from Chinese Companies in the First Quarter of 2008.

Company	Volume
Lifan	242,100
Xin Locomotive	210,100
Zong Shen	160,500
Jin Cheng	149,600
Yangtze River	141,200
Guangzhou	77,600
Prasarn	67,500
Yu Shangnian	67,200
Tongqi	66,500
Xiangbi	63,200

1.13. International Electric Bike Activities

Electric bicycles are the only vehicles at present that can achieve zero emissions. Several countries around the world are conducting R&D on and producing E2Ws. E2Ws have excellent market potential because they can offer convenience, energy savings, and an improved lifestyle. As a result of concerns over environmental protection, new lightweight materials and new technologies (such as controllers, batteries, other electronics, and electric bicycles) have reached a high state of development around the world. Electric cars, being the green transportation mode, are expected to bring huge changes to human society in the 21st century. In concert with other international technological progress, electric cars are at a breakthrough point as the Chinese auto industry enters the 21st century. Not only will China's strategic choices leapfrog development of the auto industry, they will continue sustainable development in electric bike technology. Until the present, the country's research on electric bicycles has progressed gradually. Body designs have been completed, and nickel-metal hydride (Ni-MH), Li-ion, and zinc air batteries are being developed.

Several countries around the world are involved in ETW R&D and production. Companies around the world are targeting developments for Europe, Asia, China, Taiwan, India, and Japan. ETW vehicles are considered vehicles for the common people because they help them save energy, reduce air pollution, and improve their lifestyles. They are the lowest-cost option available to the masses. This mode of transportation is a reasonable catalyst for finding socially, financially, and environmentally sound solutions to the problem of urban mobility. Table 1-9 provides statistics on the growth of bikes and motorcycles in various countries, and Table 1-10 provides data on air pollution in four Asian cities.

1.14. Electric Bike Programs in Various Countries

1.14.1. United States

Table 1-11 provides total bike sales in the United States during 2007 and 2008. Total sales in 2008 dropped 7.2% from 2007 sales. The values given are gross numbers on motorcycles (street bikes, dual-sport and off-road vehicles) and scooters. The numbers were compiled from various public sources, most of which probably trace back to the Motorcycle Industry Council (MIC), which tracks motorcycle sales in the United States.

U.S. House of Representatives (HR) 727, which was signed into law in late 2002, defines as a "bicycle" any E-bike or power-on-demand bike that has a motor of less than 750 W, an assisted speed of less than 20 miles per hour (mph), and functional pedals. No driver's license, registration, insurance, or helmet is required to operate a bicycle (in most states), and it has access to all roadways and bike paths. It must conform to Consumer Product Safety Commission (CPSC) requirements for bicycles. (Responsibility for E-bikes is assigned to CPSC rather than the U.S. Department of Transportation [DOT].) Powered two-wheelers that have larger motors, travel at higher speeds, or do not have any pedals are considered to be motorcycles in the United States; this is also covered under HR 727. In the United States, these vehicles must meet DOT requirements for motorcycles. A driver's license, registration, insurance, and helmet are normally required to operate them. Additional regulations address safety issues, brakes, lights, reflectors, and criteria for measuring the power of the motor.

The number of electric bikes in the United States is small, and the amount of information available on products is very limited. Most major bicycle companies experimented with low-power (250-W) electric bikes in the late 1990s. Consumers were disappointed with the products. Companies such as GT/Charger, Schwinn/Currie, Trek/Yamaha, Brunswick, ZAP, Ford, and Total EV/Merida offered electric bicycles in late 1990s, but most were too expensive, not powerful, and often not reliable. Most were offered through traditional bicycle dealers. Various importers of Chinese E-bikes — some well organized and with significant resources, such as Giant, EV Global, Panasonic, Currie, Ideation, Sharper Image, Pacific Cycle, ZAP, and Prima — are marketing E2Ws in the United States. The following are parameters for ETWs manufactured by TidalForce Electric Bicycle Company:

- Speed: 20–25 mph (32–40 km/h);
- Range: 20 mi (32 km);
- Motor power: 500, 750, or 1,000 W;
- Motor: direct-current (DC) brushless direct drive;
- Battery: 36-V Ni-MH;
- Power on demand;
- Fully functional bicycle; and
- Weight: 75 lb (34 kg).

Table 1-9. Growth in Asian Bicycle and Motorbike Populations, 1989–2002.

Country or Region	Annual Growth Rate (%)	2006 Bike and Motorbike Population	No. of Years It Takes for Bike Population to Double
Mainland China	25	62,105,412	3
Nepal	16	204,121	5
Vietnam	15	9,436,024	5
Philippines	14	1,735,814	5
Cambodia	13	333,663	6
Laos	11	201,948	7
India	10	41,760,670	7
Indonesia	9	16,775,380	8
Thailand	9	16,886,204	8
Bangladesh	7	285,895	10
Sri Lanka	7	710,356	10
Pakistan	7	2,729,155	11
Hong Kong, China	5	39,099	14
Taiwan, China	5	12,459,333	14
Malaysia	5	5,378,127	15
Japan	3	1,404,074	26
Singapore	1	134,799	70
Total	15	172,580,075	5

Table 1-10. Estimated Share (%) Contributed by Motorcycles to Total Transportation Pollutant Emissions in Four Asian Cities,

City	VOC	CO	PM	NO_x	CO_2
Ho Chi Minh City, Vietnam	90	70	Not available	12	40
Delhi, India	70	50	Not available	Not available	Not available
Bangkok, Thailand	70	32	4	<1	Not available
Dhaka, Bangladesh	60	26	42	4	Not available

Table 1-11. Two-Wheeler Sales in the United States in 2007 and 2008.

Vehicle Type	2007 Totals	2008 Totals	Unit Change	% Change
Dual sport vehicles	36,837	45,250	+8,413	+22.8
Off-road vehicles	209,739	146,779	−62,960	−30.0
Street bikes	647,633	611,133	−36,500	−5.6
Scooters	54,255	76,748	+22,493	+41.5
Total	948,464	879,910	−68,554	−7.2

Appendix E lists U.S. companies that market ETWs in the United States and worldwide. At present, the U.S. market is very small compared to Asian, Japanese, and European markets. U.S. companies are mainly working on the export market. The majority of companies are using lead-acid batteries, some are using Ni-MH batteries, and a small number have started using Li-ion battery technology.

1.14.2. Japan

Motorcycle production in December 2008 was recorded as 112,024 units, or 10.6% less than the total recorded for December 2007. December 2008 represented the 16th consecutive month of production decreases. Motorcycle production for calendar year 2008 (January–December) was recorded as 1,226,839 units, or 26.8% less than the total for 2007. Year 2008 represented the third consecutive year of annual production decreases. Table 1-12 shows motorcycle production by manufacturer.

The Japanese Automobile Manufacturers Association (JAMA) reported that a number of negative factors affected Japan's motorcycle market in 2008, including higher vehicle prices (reflecting mandatory compliance with new emission regulations) and a deteriorating economic environment (stemming from the global financial crisis). Demand also declined in Japan as a result of stronger crackdowns on illegal parking and the chronic shortage of motorcycle parking bays in cities and towns. On the other hand, surging fuel prices underscored the economy and convenience of electric bikes to consumers.

In 2009, manufacturers are expected ti sell about 400,000 electric bikes in Japan; almost all will be made by Japanese companies, such as Yamaha, Panasonic, Nissan, and PUES Corporation. Because Japanese regulations are unique to Japan and relatively expensive to conform to, it is difficult for outside companies to sell electric bikes in Japan. There is little or no opportunity for Chinese companies in this market.

The Nissan Pivo 2 concept E-bike was selected as one of the 10 "coolest" big electronic products by InfoWorld in Japan. Nissan displayed the Pivo 2 concept E-bike at the Tokyo auto show. This kind of ETW can be folded and its wheels can be turned 180 degrees, which make it convenient to park and lock. It also has a speech recognition system, so the driver can ask it questions, such as where it is parked. Its facial recognition function can distinguish the driver's face from others to identify the bike's owner. This kind of vehicle will be available in the market around 2015.

The Japanese Koga Miyata Industrial Corporation developed an ultra-featherweight electric bicycle, weighing only 16.8 kg. This bicycle went on the market recently. It uses Li-ion battery technology. It also has a failure diagnosis intelligent system.

Table 1-12. Japanese Motorcycle Production by Manufacturer in 2008.

Motorcycle	December 2008		January-December 2008	
	No. of Units	Change from Previous Year	No. of Units	Change from Previous Year
Honda	28,586	101.7	337,339	72.4
Suzuki	31,816	86.2	348,465	81.0
Yamaha	32,165	78.8	349,604	69.8
Kawasaki	19,455	100.3	191,337	68.7
Others	2	40.0	94	27.3
Total	112,024	89.4	1,226,839	73.2

Source: Japanese Automobile Manufacturers Association (JAMA), January 2009.

PUES Corporation is marketing advanced electric motor bikes with Li-ion battery technology. They have a range of up to 70 km. During 2008, 1,500 of these motorbikes were sold.

Yamaha has a long history of electric scooter development. It started a program in 1990 to develop a "clean, silent" scooter, then developed an electric scooter with an advanced network control system. In 2003, it went on to develop a "light, clean, smart, silent" commuter scooter with Ni-MH battery technology. In 2007, it developed a full-performance commuter electric scooter that uses Li-ion battery technology with a 48-V system and has a 60-km range. Yamaha bikes are quite popular in Japan. Its ETW models — the PAS Li S, PAS CITY-S, and PAS CITY-F — came on the Japanese market recently. They cost $1,200 and include Li-ion battery packs. The battery has an estimated range of 39 km in standard mode, 21 km in power mode, and 67 km in auto-eco-plus mode.

Panasonic is one of Japan's largest electric bike manufacturers. Its E-bike is a good solution for people who want an electric folding bike. It is lightweight, foldable, comfortable, and can easily go up any hill. It has an Li-ion battery and advanced motor and controller system.

KTM is bringing an electric bike to the market by 2010. Quantya already has an electric dirt bike — the Strada — available in the market. Honda, while quietly leading the way in the bike industry, is proclaiming its electric future loudly. The latest addition to the electric bike stable is Honda's new Vectrix Superbike concept, which was revealed in Milan. This model uses Li-ion battery technology along with an advanced motor and control system. It also has a smart mode that allows the rider to locate the bike by using a voice recognition system. Honda is partnering with Yuasa to add an electric bike to the 2012 lineup. Yuasa and Honda will develop the new Li-ion batteries. They plan on building a manufacturing facility near Kyoto in Japan, at an estimated cost of $18.5 million.

Toshiba has revealed its Cannondale E-bike. Following in the footsteps of Yamaha and Panasonic, whose bikes were unveiled in July 2008, Toshiba exhibited a model manufactured by Cannondale during the 15th World Congress on Intelligent Transport Systems. The bike is equipped with a battery module that is based on Toshiba's proprietary quick-charging Li-ion rechargeable battery, "SciB." Measuring 100 x 300 x 45 mm, the module features a 24-V battery and weighs 2 kg. The bike has an assisted travel range of up to 30 mi on a single charge. The bike is scheduled for release in America and Europe early in 2010, with Cannondale reportedly planning to price it at about $3,000.

1.14.3. European Union

Europeans will purchase about 187,000 electric bikes in 2009. Most will be European brands that are made or assembled in European countries but will have components from Taiwan, Japan, and China. The number of E-bikes in Europe is estimated to grow by more than 1 million every year. In countries like the Netherlands, Germany, and Switzerland, Europeans are willing to pay high prices for electric bikes with high quality and performance. In other countries, such as Italy and the Eastern European countries, low-priced vehicles are needed.

Electric bicycles need a 250-W motor with a sensor and controller that require the rider to pedal. Larger motors and power-on-demand can be sold if the buyer has the correct approval for the type of vehicle desired. Safety and quality issues are important; two-year warranties are required by law.

The European Parliament updated its motorcycle and moped classification. It now considers some electric pedal-assisted bicycles (EPACs) as a separate class, a move that affects the legal requirements for their riders. Owners of EPAC that fall into this new classification no longer need to conform to helmet, licensing, and other laws that apply to motorcycles. The Parliament defined EPACs as electric bikes that apply power assistance when the pedals are pushed only when speeds are 15.5 mph and lower. At speeds higher than that, or when the rider stops pedaling, power assistance must be cut off. Electric bikes that meet the standard are treated like bicycles.

1.14.4. Germany

Germany is a bicycle country. Every year, the volume of bicycle sales reaches several million units. The entire bicycle sales market is very large. Germans choose to ride bicycles for several reasons. According to German statistics, 3.7% of German adults ride bicycles for several years after they have retired. They ride bikes for fun and leisure. Middle-aged Germans ride bicycles for exercise. They use bikes for short-distance journeys, shopping trips, and to go to and from work. However, at the present, the volume of German electric bicycle sales is about 1% of the volume of automobile sales in Germany, or about 45,000 ETWs. The average price of an electric bicycle is 1,000 Euros. If the market share increases by 1%, it will increase the market potential by 45,000,000 Euros. This represents a very large market. Germany has extensive laws to regulate its bicycle industry. The text that follows provides information on some German bike manufacturers:

German electronics manufacturer Heinzmann opened a new E-bike and electrocycle retail store in the town of Shonau, Germany, in the Black Forest near the Swiss border. In the town, Heinzmann also operates a facility that manufacturers its hub motors and electric bicycles. The hub motors are also used in E-bikes made by Los Angeles-based EV Global Motors.

Ecobrand Exim International Company, Ltd., manufactures and exports a wide range of electronic and E-bikes, from basic E-bikes with 24-in. wheels and 1 50-W side motors to the topof-the-line Harrier Sport Rider DL electronic mountain bike with 36-V, 200-W motor, carbon fiber wheels, derailleur gears, and front and rear disc brakes.

1.14.5. Switzerland

Bicycles are very popular in Switzerland. People there are very conscientious and want to use clean transport vehicles. The Swiss national program NewRide was developed to enhance the market introduction of ETWs. Its main activities are to strengthen the commitment of manufacturers and importers, improve the competence of the local NewRide dealers, and create local networks to provide communication platforms as part of introducing ETWs in the market. In 2002, nine Swiss cities participated. The main objective of NewRide is to promote ETWs by improving market conditions for the suppliers (manufacturers, importers, and dealers).

The vehicles in the NewRide program consist of nine brands of ETWs with a total of more than 20 models. NewRide collaborates closely with vehicle manufacturers and importers, who make commitments to better train dealers. Furthermore, they provide vehicles for NewRide exhibitions and participate in joint advertising. The main information tool is the web site www.newride.ch, which contains the following: general news, description of the NewRide program, NewRide cities, a vehicle catalog, NewRide dealers, NewRide companies, frequently asked questions, events, media, contacts, and the NewRide-dealer label.

In spring 2002, NewRide introduced the NewRide-dealer label. Acquiring this label obliges the dealer to obtain the skills required to provide professional advice and offer aftersales service. The label is a vignette that is put on the shop window. The dealers benefit from the training offered by NewRide, the promotion campaigns, and the exchange of experiences. The NewRide Program is organizing ride and drive rallies for citizens as part of trade fairs or as separate events to make people aware of the benefits of electric bikes. The events also serve as platforms for the dealers, who can benefit not only from the event itself but also from the publicity they get by joint advertising. They benefit for two reasons:

- *Promotion is community based.* Intensive promotion by local authorities, which is supported by NewRide program management, increases people's awareness of and — because of the program's governmental background — their confidence in new products. Although subsidies are not a main issue in the NewRide program, NewRide cities are free to help promote the introduction of ETWs into the market by offering financial incentives. For example, Basel and Zurich offer a subsidy of 10% and 20%, respectively, on the ETW purchase price.
- *The program is highly appreciated.* NewRide's approach of improving the conditions for the market introduction of ETWs according to the means and skills of the partners seems to be promising. The program is highly appreciated by local authorities as well as the media and general public. The final breakthrough of ETWs in terms of sales has not been achieved yet, however. With only 7,000,000 residents, Switzerland is probably too small for a large market introduction of ETWs. To reach a level of market penetration that would justify industrial production, an internationally coordinated promotion is needed. There is only one ETW manufacturer in Switzerland — SwissLEM A.G. — that manufactures and distributes electric cars, electric motorcycles, and electric bicycles, including the "Twike."

1.14.6. India

India is the largest democratic country in the world, with a population of about 1.2 billion. People there generally use three-wheeled vehicles to commute from home to railway stations and bus depots, to shop, to visit friends in town, and to make short visits to supermarkets. These vehicles are always available for hire for a fee. During a given day, a three- wheeler typically travels 40 to 50 km. Two-wheelers are used for the same purposes as three- wheelers and are also used for going to and from work. Table 1-13 shows the number of both types of vehicles in India. Figure 1-2 shows the percentages of the various vehicles that are on the roads in India.

Two- and three-wheeled vehicles are seen as the most potent zero-emissions option for local vehicles in the near future. Local companies are developing some of them now: (1) an electric auto-rickshaw is undergoing user trials and (2) an ETW is in the prototype stage. So far, there has been little market for E-bikes in India, but they are very appropriate for the uses mentioned and should be popular as they become available and as incomes improve. Price is an important issue. The largest market after China will eventually be India.

In an effort to keep costs down, Ace Company is planning to set up a facility for manufacturing advanced battery technologies. It is working with a Taiwanese or Korean manufacturer. Although it is targeting young people, sales to them are unlikely, but institutional sales should be good. According to the Society of Indian Automobile Manufacturers (SIAM), EVs attract an 8% excise duty and earn institutional buyers a sizable subsidy. Table 1-14 provides a comparison.

Hero Cycles and others are already targeting such business-to-business sales; they want to sell E-bikes to large businesses to use for short trips, which saves energy. Companies and organizations whose biggest cost is the delivery of their products have shown interest in buying bulk E-bikes. The strategy, at least for Hero, is to develop advanced ETW technologies and bring their cost down to attract large businesses to become consumers.

Indeed, electric cycles fitted with motors are fairly common and can be charged by being plugged in. In countries like China and the United States, electric bikes are considered the "in" thing. Concerns over global warming have also pushed Western countries to come up with regulations to promote electric bikes.

There has been a move to Ni-MH and Li-ion batteries in Japan, Europe, and the United States. This same progression will occur in China and India as battery volumes grow and prices go down. The advantages of Ni-MH and Li-ion batteries are their lower weight and size and longer range if manufacturers install more battery capacity. ETWs may be less expensive to run, but they have other problems. Because ETWs are lightweight, maintenance becomes a big issue on bad roads.

1.14.6.1. Bajaj Auto

Bajaj Auto is currently working on electric two- and three wheelers as follows:

Table 1-13. Number of Two- and Three-Wheeled Vehicles in India, 2002–2008.

Category	2002–03	2003–04	2004–05	2005–06	2006–07	2007–08
Three-wheeler	231,529	284,078	307,862	359,920	403,910	364,703
Two-wheeler	4,812,126	5,364,249	6,209,765	7,052,391	7,872,334	7,248,589

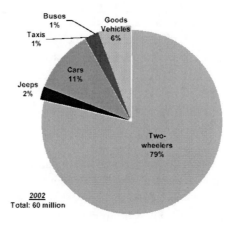

Figure 1-2. Percentages of Different Types of Vehicles on the Road in India.

1. Electric three-wheeled auto-rickshaw program
   Flooded lead-acid batteries
   Direct drive to rear wheel through a transaxle
   Range of ~100 km in city running conditions
   Four vehicles undergoing field trials in Agra
2. Electric two-wheeled scooter program
   Motor mounted on rear wheel
   Sealed gel-type lead acid batteries
   Expected range of ~ 65 km in city running conditions

Bajaj has an R&D program on advanced battery and motor and controller technologies.

Advanced Battery Technology for Electric Two-Wheelers in the People's ...

Table 1-14. Comparison of E-Bikes and E-Scooter Models in India.

Brand Name[a]	Motor Power (W)	Range (km)	Recommended Load Capacity (kg)	Operating Voltage (V)	Approx. Price Range (Rs)	Wheel Size (in.)	Special Features
Hero Ultra Advanta[a]	<250	45	75	36	15,000–20,000	16×2.125	Handle bars, indicator lights, color finish
Hero Ultra Extra	<250	60	75	36	15,000–20,000	16	Body color, rear view mirrors, turn indicators, under cover protection, useful utility box
Hero Ultra Maxi[a]	<250	50	90	48	29,000	16 3	Body line, color finish, easy-to-read instrument panel, indicator lights
Hero Ultra Velociti	500	50	NA[b]	48	35,225	10	Hydraulic disc brakes in front, drum brakes in rear, telescopic suspension
Yo Speed[a]	750	75–80 at 70 kg, 55–60 at 135 kg	130	NA	NA	3×10	High-speed 750-W motor
Yo Spin	<250	55	75	NA	NA	18× 2.125	Efficiency, style, performance
Yo Smart	<250	75	75	NA	NA	16×2.5	Comfort
Yo Trend	<200	50	75	NA	NA	22	Strong body frame and clear headlight lens for control and safety
Yo Tuff	<200	50	75	NA	NA	26	Body line, comfortable seat, braking, headlight
Yo Trust[a]	<200	75	75	NA	NA	24	Effective style, good performance

[a] For more information, see http://www.surfindia.com/automobile/hero-ultra-advanta.html, http://www.surfindia.com/hero-ultra-maxi.html, http://www.surfindia.com/yo-speed.html, and http://www.surfindia.com/yo-trust.html.

[b] NA means not available.

1.14.6.2. Leo Bikes

Leo Bikes is an automobile division of Sagar Impex, an exporter, importer, and manufacturer of textile machinery and many other types of products. Leo Bikes is dedicated to developing and commercializing battery-operated vehicle (BOV) technologies, with an initial focus on two-wheeled applications. It believes that an increase in personal traffic transportation should not be at the cost of a life-sustaining environment and that valuable natural resources must be saved for the next generation. It is using the latest technology and manpower to maintain the world-class quality of its products

- Business type: Manufacturer, wholesale supplier, exporter, importer
- Products: EVs, electric scooters, electric bicycles
- Address: 26-Ground Floor, Rajhans Point, Varacha Road, Surat, Gujarat, India 395006
- Telephone: 91 0261 2501312

- Fax: 91 0261 3073412
- Web site: http://www.leoebikes.com
- E-mail: send to Leo Bikes

1.14.6.3. Accura Bikes Pvt., Ltd.

- Business type: Manufacturer, retail sales, wholesale supplier, exporter, importer
- Products: Electric bicycles, electric cars, and batteries for EVs
- Services: Design, research, maintenance, repair
- Address: E-7/3 Upsidc Industrial Area, Selaqui, Dehradun, Uttranchal, India 248197
- Telephone: 0135-6450412
- Fax: 0135-2698832

1.14.6.4. Birionic Pvt., Ltd.

- Business type: Manufacturer, retail sales
- Products: Air-cooling systems, electric bicycles, light-emitting diode (LED) lighting, small wind energy systems
- Address: 3F Palmtree Place, Palmgrove Road, Victoria Layout, Bangalore, Karnataka India 560047
- Telephone: 91 8051 128155
- Fax: 91 8025 555986

1.14.6.5. Callidai Motor Works Pvt., Ltd.

- Business type: Manufacturer, exporter
- Products: Electric (battery-powered) bicycles, wheelchairs, and tricycles. It can design and manufacture any small battery-powered vehicle. Products could include battery-powered vehicles to move people and goods for use in airports, railway stations, and industry, among others. It makes its own motor- speed controller by using pulse-with-modulator (PWM) and battery chargers
- Services: Consulting, design, project development, research
- Address: 28 Desika Road, Mylapore, Chennai, Tamilnadu, India 600004
- Telephone: 91 4449 91609
- Fax: 91 4449 95185

1.14.6.6. Radha Energy Cell

- Business type: Manufacturer, retail sales, wholesale supplier
- Products: DC-to-AC power inverters, electric bicycles, photovoltaic cells, solar collectors, solar traffic lights, solar street lights, solar lights, solar water heaters, solar inverters, solar cells
- Services: Consulting, installation, manufacturing

- Address: First Floor, Deol Market, Rajesh Nager, Badi Haibowal, Ludhiana, Punjab, India 141001
- Telephone: 09 8888 97248 and 09 8150 97248

1.14.6.7. Rashron Energy and Auto, Ltd.

- Business type: Manufacturer, exporter, research institution
- Products: Electric bicycles, solar cooking systems, large wind turbines
- Address: 603 GIDC, Makarpura, Vadodara, Guj, India 390010
- Telephone: 0265 643224
- Fax: 0265 643778

1.14.6.8. Rotary Electronics Pvt., Ltd.

- Business type: Manufacturer, exporter, importer
- Products: Electric scooters, electric bicycles
- Services: Engineering
- Address: 18 Fifth Cross, Fourth Main, Indl. Town, WOC, Rajajinagar, Bangalore, Karnataka, India 560044
- Telephone: 91 8023 353662
- Fax: 91 8023 380183

1.14.7. Taiwan

Taiwan was the first country in the world to implement a zero-emissions two-wheeled vehicle mandate. To support the government policy, the Industrial Technology Research Institute (ITRI) developed two generations of electric scooters (ESs) by implementing VRLA and Ni-MH batteries, respectively. The first-generation ES, which was adopted by Sam-Ever Company, began production in September 1999. Today there are more than 30,000 ESs on the road in Taiwan; however, most ES users are not satisfied with the cruising range, vehicle weight, charging time, and vehicle cost. To meet market requirements and support the government's mandatory ES sales regulation, ITRI is improving the vehicle's overall performance. It is developing the next-generation ES by incorporating Li-ion battery technology and higher-power electronics, motors, and control systems. The engineering targets are to improve current performance by 25% in terms of weight, 50% in terms of cruise range, 20% in terms of total energy efficiency, and 300% in terms of battery life, with no price increases after the government subsidy. According to a customer survey disseminated by the government, the ES being developed will satisfy most current ES users. With these improvements, it has excellent potential for replacing most 50-cc gasoline scooters and becoming one of the highest-class scooters in Taiwan. Its structure has been redesigned by replacing the steel welding frame with a new aluminum casting frame. The Li-ion battery has a management system to optimize and protect the battery cells. The vehicle performance has been upgraded by optimizing the phase angle and weakening the flux of the motor and controller to increase the torque in both low- and high-speed regions. A single-stage timing-belt transmission will replace the two-stage gear transmission. The centralized vehicle energy-management system will monitor and control the battery inputs and outputs to increase the vehicle's total efficiency. Finally, a system bench test and a vehicle dynamometer test will validate all engineering targets.

As a result of the significantly improved cost performance of the Li-ion batteries and the core competency of the LEV industry, a series of LEVs, including electric bikes and scooters and advanced-performance personal vehicles, was introduced in the global marketplace by Taiwanese industries, as well as by ITRI.

To improve transportation efficiency and environmental quality, a new national campaign for the LEV program will be launched in 2009. It will focus on 800-W, lightweight, sit-down, two-wheeled ESs and 30-kW lightweight electric cars. The government strategy is to develop a policy that will cover innovative technologies. It will address (1) cost-effective and safer battery and power electronics, (2) forming a technological platform for industry to develop a cluster of key components, (3) standards for testing and certifying the key components and the vehicle, (4) a business model for establishing the infrastructure, and (5) battery and vehicle operation for promotion in the LEV national program. The parameters for developing an ES are given in the Table 1-15.

Table 1-15. Parameters for Developing a Small, Light Scooter.

Aspect	Parameters
Dimensions	Length of <2.5 m, width of <1 m, height of <2 m
Stand stability and reliability	Stability test of side stand and middle stand 10,000 times durability kickoff test for each stand 15,000 times durability kickoff test for vehicle equipped with only one stand
Weight	<60 kg
Normal voltage	<48 V
Motor power output	<1 kW
Speed	<30 km/h
Tires	4.10/3.50 5
Frame fatigue strength	Under a specified load, about 6.6–10 Hz; 2 G (force from gravity) after 70,000 cycles of vibration, no rupture
License plate	Required
Others	Front and rear lights, horn, turn and brake signals, license plate light, rear refractive sign

Table 1-16. Parameters for Developing Two-Wheeled Light Electric Vehicles.

Parameter	LEV A	LEV B	LEV C
Market segment	Short range, charge infrastructure required, near-term products, supersede 10% of the 50-cc market	Medium range, charge infrastructure not required, mid-term products, supersede 30% of the 50-cc market	Longer range, long-term products, technology innovations, supersede the 50-cc market
Weight (kg)	<50	<70	<90
Speed (km/h)	20–35	35–45	50
Range (km)	15–25	25–35	>50
Battery	Lead acid	Ni-MH and Li-ion	Li-ion and fuel cell
Price compared to that of 50-cc scooter	Less	Equal	More

At present, there are 10 million motorcycles on the road in Taiwan, and many environmental issues need to be addressed in most of the metropolitan areas. For the past 15 years, the Taiwan Environmental Protection Agency (EPA) has implemented a series of emission regulations that are so stringent that two-stroke engine scooters can hardly pass them. These kinds of regulations will allow LEVs to penetrate the market in Taiwan and improve air quality. The government mandated that 2% of ESs be integrated in the total number of automobiles in Taiwan. According to the Taiwan EPA, motorcycles still contribute 35% to carbon monoxide (CO) emissions and 18% to total hydrocarbon emissions.

In 2006, a 5-year national ES project was established by the Taiwan EPA to help the industry sell half a million electric scooters. By 2012, their use is expected to reduce CO emissions by 7,000 tons, hydrocarbon and NO_x emissions by 4,000 tons, and CO2 emissions by 56,000 tons each year. The development of these LEVs is shown in Table 1-16.

1.14.7.1. Industrial Technology Research Institute

The Industrial Technology Research Institute and two companies are involved in developing and producing E-bikes. ITRI is a nonprofit R&D organization with 6,000 employees engaged in five areas of applied research and industrial services:

1. Information and communication,
2. Material and chemical technologies and nanotechnologies (including Materials Research Laboratories, which have a high-tech facility for developing and building Li battery systems),
3. Biomedical technologies,
4. Advanced manufacturing and systems (including Mechanical and Systems Research Laboratories, which develop clean power systems, vehicle electronics, advanced vehicle systems, automobile engines, and hybrid propulsion systems), and
5. Energy and the environment.

ITRI's Mechanical and Systems Research Laboratories developed an electric scooter that uses a 2-kWh Li-ion battery back developed by the Materials Research Laboratories. The curb weight is 100 kg and the range is 60–75 km (city mode). The 3-kW brushless hub-wheel motor allows a top speed of 60 km/h. ITRI is expecting a reduction in the price of Li batteries in the next few years. The scooter will be ready for mass production within two years.

ITRI's Mechanical and Systems Research Laboratories were also involved in developing an ES that uses two types of batteries: a lead acid battery for propelling the motor and a Li-ion battery for the energy supply. The latter battery can be removed and charged at any household outlet. The batteries weigh 2.5 kg total. Both provide a range of about 30 km. Because most of the people who live in towns live in apartments without garages, there is a lack of charging stations; this lack needs to be addressed in a cost-effective way.

1.14.7.2. KOC Industry Corp.

- Business type: Manufacturer, exporter
- Products: Electric bicycles
- Address: 5-4 Kaifa Road, Nantze Export Processing Zone, Kaohsiung, Taiwan 811
- Telephone: 886-7-3651147
- Fax: 886-7-3657834

1.14.7.3. Shihlin Electric and Engineering Corporation (a division of KOC)

- Business type: Manufacturer, wholesale supplier
- Products: Electric lawn mowers, electric bicycle components, EV components, electric bicycles, electric scooters, air heating system components, DC motors, automotive electrical components, electrical contractor supplies
- Address: 12F, 90, Sec. 6, Chung-Shan N. Road, Taipei, Taipei County, Taiwan 100
- Telephone: 886-2-2834-2662, extension 149
- Fax: 886-2-2835-9250

2. LITHIUM-ION BATTERY TECHNOLOGY FOR ELECTRIC TWO-WHEELERS

Li-ion battery technology is widely used in portable electronic products, such as cell phones, camcorders, and portable computers. At present, the technology is gaining worldwide attention as an option for transportation applications, including EVs, hybrid vehicles, PHEVs, fuel cell vehicles, and electric bikes in Asia. The United States continues to lead in Li-ion technology R&D, where a strong R&D program is being funded by DOE and other federal agencies, such as the National Institute of Standards and Technology and the U.S. Department of Defense. In Asia, countries like China, Korea, and Japan are commercializing and producing this technology. China has more than 120 companies involved in the production of Li-ion battery technology.

Mass manufacturers of Li-ion cells for consumer products are now engaged in the developing Li-ion chemistries for hybrid vehicle, electric bike, and PHEV applications, with commercialization possible as early as 2012. The major impediment to engaging in the development of Li-ion batteries for PHEVs appears to be the fact that the requirements for PHEV batteries are not sufficiently defined at this time. The apparent interest of General Motors Corporation (GM) in PHEVs might stimulate efforts to develop Li-ion battery technologies for PHEV applications. Several companies in Europe and Japan have been developing medium- and high-energy Li-ion technologies, some of them based on advanced materials, chemistries, and manufacturing techniques. Their strategy is to pursue limited-volume applications and markets that may be emerging, especially in small battery-powered EVs, electric bikes, and, more recently, PHEVs. Several of these companies hold the view that PHEVs powered by Li-ion batteries and small battery-powered EVs will be able to match the life-cycle cost- competitiveness of conventional vehicles in urban fleet applications, and a few have established cell-production capacities for hundreds to a few thousand of 10–25-kWh batteries per year, which may be sufficient for demonstration fleets.

2.1. Battery Manufacturers and Performance

The demand for Li-ion rechargeable batteries has been driven by the rapid growth of portable electronic equipment, such as cell phones, laptops, and digital cameras. In addition, the expectation that rechargeable batteries will play a large role in alternative energy technologies, as well as in E-bikes, EVs, hybrid vehicles, and PHEVs, has made the development of Li-ion rechargeable batteries a fast-growing industry in China and the world. The first commercial Li-ion rechargeable battery, which was introduced by Sony Japan in 1989, used graphite as the anode. Since then, Chinese companies have been developing and producing Li-ion batteries for portable applications. Recently, the number of Chinese companies that manufacture Li-ion batteries has grown by several hundred. Large companies such as BYD Battery Co., Ltd. (BYD); Shenzhen BAK Battery Co., Ltd. (BAK); Shenzhen B&K Technology Co., Ltd. (B&K); and Tianjin Lishen Battery Joint-Stock Co., Ltd., have a large share of the global market. The performance characteristics of Li-ion cells from various manufacturers are given in Table 2-1.

Table 2-1. Performance Characteristics of Li-Ion Technology for E-Bike Applications.

Manufacturer	Cell/Module	Voltage (V)	Capacity (A·h)	Weight (kg)	Specific Energy (Wh/kg)	Power Density (W/kg)	Cycle Life	Application
MGL	M	24	13	2.3	92	151	500	Bike
Beijing Green Power	C	3.8	10	0.110	118	84	540	Bike
Tianjin Lishen	C	3.6	13	0.360	117	91	500	Bike
Battery Joint-Stock Co., Ltd.	C	3.6	8	0.245	121	94	500	Bike
Tianjin Lantian	C	3.6	2	0.044	NA[a]	NA	NA	Small bike
	C	3.6	18	0.80	115	100	800	EV/bike
	C	3.6	100	2.6	106	138	460	EV/bike
	M	24	20	10.0	52	53	400	EV/motor bike
Xingheng	M	NA	10	0.37	100	200	500	EV/bike
Suzhou Phylion Battery Co., Ltd.	C	3.7	10	0.36	102	86	500	E-bike
Shenzhen BAK Battery Co., Ltd.	C	3.7	1.8	0.054	123	100	NA	E-bike
ABT, Inc.	C	3.7	5	0.270	69	76	>700	E-bike
GBP Battery Co.	C	3.7	60	1.80	123	82	>550	EV/bike
Hyper Power Co.	C	3.7	1.15	0.037	115	78	>600	Bike
Shenzhen B&K Technology Co., Ltd.	C	3.8	0.100	0.007	100	NA	400	Bike
Tianjin Blue Sky	C	3.7	60	1.8	122	80	>500	EV/bike
BYD Company, Ltd.	C	3.7	1.8	0.046	144	98	>400	E-bike
EMB Battery Co.	C	3.7	2.1	0.045	172	88	300	E-Bike

[a] NA means not available.

Table 2-2. Performance Requirements for Chinese E-Bikes.

Parameter	Value
Speed	25–40 km/h
Power	240–500 W
Voltage	36–48 V
Range	40–75 km on single charge
Fuel efficiency	70–80 km/kWh
Specific energy[a]	110 Wh/kg
Energy density[a]	170 Wh/L
Power density[a]	350 W/L
Cost per kWh[a]	$505
Cycle life[a]	900 yr

[a] Li-ion technology.

2.2. Electric Two-Wheeler Performance Requirements

ETW performance is critical to customers. To Chinese riders, speed is important. Table 2-2 lists performance requirements for Chinese electric bikes.

2.3. Safety

Safety is a primary concern of Chinese government officials. In the 2003–2005 period, there were more than 100,000 road fatalities in China, and most of the victims were vulnerable road users, such as pedestrians or bicyclists (National Bureau of Statistics 2005 a). One of the motives cited for regulating the use of gasoline-powered motorcycles is safety. Beijing officials cited safety as one of the main reasons to ban electric bikes as well. The China Bicycle Association (electric bike advocates) countered, citing the crash rate (percent of vehicles involved in a crash per year) for electric bicycles is only 0.17% and is 1.6% for cars. The primary question is "Do electric bicycles decrease the safety of the entire transportation network in terms of the number of fatalities and injuries per person per kilometer traveled? That is, is the incidence of fatalities higher for electric bike users because they are vulnerable road users?" To address safety concerns, the operating speeds of electric bikes have been limited so that they can safely operate in bicycle lanes. Another question is "If one assumes that a traveler will take a trip regardless of the transport mode, what are the safety implications of switching to an alternative mode (bicycle or transit)?"

2.4. Battery Industry for Two-Wheelers

The PRC is the most highly populated country in the world. It has been involved in battery technology developments and manufacturing for several years. Over the last 10 years, it has exported the largest number of batteries for telecommunication, computers, cell phones, and other electronic equipment to many countries. Several hundred Chinese companies, small and large, are involved in developing lead acid, Ni-MH, and Li-ion batteries for these applications. In the past three to four years, companies outside the PRC have been bringing advanced battery technologies to the PRC. They have been setting up partnerships and joint ventures to manufacture batteries for these and other applications (such as E-bikes, EVs, and HEVs) to take advantage of the low labor costs in China and the incentives provided by the Chinese government. Companies in the PRC are very aggressive about developing manufacturing processes for the battery export market.

In 2003, the annual mobile phone production capability of Chinese telecommunication equipment manufacturers reached 200 million sets. Actual annual output was 186 million sets, of which about 120 million sets were exported. That accounted for one-quarter of the total global output of mobile phones. Driven by the mobile phone market, the telecommunication equipment manufacturers improved their share of the export market. Chinese companies produced 334 million batteries for mobile phones alone in 2003.

Taking advantage of small early markets for Li-ion battery technology, Chinese companies were involved in developing a large number of advanced batteries with foreign companies. Now these companies are developing Li-ion batteries for E-bikes, EVs, HEVs, and PHEVs. From 2001 to 2004, the number of battery companies in China increased from 455 to 613. Accordingly, the number of employees also increased, from 140,000 in 2001 to 250,000 in 2004. The total output reached 63.416 billion ¥ ($8.1 billion) in 2004, an increase of about 53% over 2001 output.

The sales of batteries produced by large-scale companies in the battery industry were 59.82 billion ¥ ($7.65 billion) in 2004 — an increase of 53% over sales in 2003, an increase of 105% over sales in 2002, and an increase of 161% over sales in 2001. This growth is attributed to the growth of large companies. In the period 2001–2004, the debt-to-asset ratio of China's battery industry fluctuated between 54% and 59%.

2.5. International Competition

The Li-ion battery was commercialized by Sony Company in 1992. From then until 2001, 95% of the Li-ion battery market was dominated by Japan. In more recent years, the Ni-MH and nickel-cadmium (Ni-Cd) battery markets have been gradually shrinking. The cell phone, digital camera, portable digital assistant (PDA), portable video camera, and other portable device industries have been getting stronger. The Li-ion battery is showing great potential for these devices, as well as for E-bike, EV, HEV, and PHEV applications. These applications are driving the development of Li-ion battery companies in China, such as BYD and TCL Hypower, and in Korea, such as Samsung SDI and LG Chemicals, with the help of the Korean consumer electronics industry. So far, the world market for Li-ion battery technology is being shared by China, Japan, and Korea.

Since 2001, China's Li-ion battery industry has been growing rapidly as Shenzhen BYD, B&K, BAK, Tianjin Lishen, and others have been conducting R&D for various applications, particularly ETW and electronics applications. The average yearly growth has been more than 140%. In 2002, China produced 270 million Li-ion battery cells, which took 20% of world market. So far, China trails Japan as the second-largest producer of Li-ion batteries in the world. The competitive capability of China's Li-ion battery industry is summarized in the three points that follow:

- Compared with Korea and Japan, China has three advantages: (1) low production cost with abundant labor resources, (2) the largest consumer market, and (3) a relatively complete supply chain for the Li-ion battery.
- With regard to disadvantages, Chinese companies are still weak in R&D of core technologies. The top technologies associated with Li-ion batteries are still controlled by developed countries like the United States, Japan, France, and Canada.
- China's enterprises are relatively small and lack experience and expansion in the international market.

Korea started the secondary battery business relatively late, but Korean companies benefited from the strong growth of information technology (IT) terminal products for consumer and portable devices. In the field of cell phones and two-dimensional monitors and disc players, Korea has excellent brands, like Samsung and LG. So far, in Korea, Samsung SDI, LG Chemicals, and SKC are Li-ion battery enterprises that could be internationally competitive. However, Korea depends on imports for 80% of its raw materials and other parts for Li-ion batteries, which could weaken its ability to compete over the long-term. The Korean government is pushing to form a complete supply chain for the Li-ion battery industry.

High cost is the main reason for the decrease in the competitive capability of Japanese companies and the decrease in their presence in the market. But Japan still has an obvious advantage over China and Korea with regard to core technology. Japanese enterprises still own the market for high-end electronic products because of this core technology advantage; the cost of their Li-ion battery products is as much as three times higher than the cost of China's products.

However, with the rapid growth of the Li-ion battery industry, Japanese Li-ion battery companies have been forced to confront challenges by taking measures and changing strategies to improve their performance and reduce their costs. For instance, Sanyo and Sony are putting pressure on China's companies with regard to intellectual property; Japanese enterprises are building up production bases in Beijing and Wuxi to take advantage of cheap Chinese labor; and Japanese companies like Sanyo and Sony expanded their production base to keep their leading position, while ATB and NEC are trying to reduce their costs and strengthen their R&D.

Japanese enterprises are still playing a leading role in Li-ion battery technologies with high levels of automation. Both China and Korea are importing automated technologies and equipment from Japan. Although China and Korea have been gaining a bigger market share with their low-cost products, Japan had 56% of the world market in Li-ion battery in 2007. Korea had 15%, an increase from 12% in 2006. China had 29%, a decrease of 3% from the share in 2006.

2.6. Policy Perspective

Most all of the countries in the world are paying more attention to green energy industry. The EU plans on one-fifth of its energy use to come from renewable energy sources by 2012. Germany is investing 2 billion Euros to explore wind energy. China is developing a green energy industry to realize sustainable development. A key point of China's national energy plan is the strategic development of new energy and renewable energy in 2000–2015, by promoting the exploration of wind, solar, geothermal, biomass, and other energy sources. By 2015, China's new energy and renewable energy should amount to 2% of total energy consumption.

China is also supporting the secondary battery industry, such as Li-ion batteries that have a high energy density and long cycle life. The Ministry of Science and Technology included advanced batteries and their materials in its Ninth 5-Year Plan and EVs in its Tenth 5-Year Plan. Policies that incorporate incentives for new types of secondary batteries include these, among others:

- "National Key Plan for Development of High-Tech," "National Demonstration Engineering for High-Tech Industrialization," and "National Creative Plan of Technology" of the National Economic and Trading Committee;
- "863 Plan for High-Tech Research and Development," "973 Plan," "Torch Plan," "Star-Fire Plan," "National Plan for Key New Products," and "Creative Foundation for Medium and Small Business" of the Ministry of Science and Technology; and
- "Electronic Development Foundation" of the Ministry of Information Industry.

2.7. Trends in Lithium-Ion Battery Development

Japan is the largest producer of Li-ion batteries in the world and owns most of the patents related to Li-ion batteries. The Li-ion battery industry in China started later but developed very rapidly. Statistics (incomplete) indicate that about 200 companies in China produce Li-ion batteries and related materials. It is predicted that China will soon be the biggest producer of Li- ion batteries, beating Japan.

So far, most of a Li-ion secondary battery is made of $LiCoO_2$ as the cathode material and carbon as the anode material. The process for making it is quite mature, and the production cost is actually lower than that for a Ni-MH secondary battery. $LiCoO_2$-based Li-ion batteries have a safety issue, however; they could cause an explosion, especially under some extreme conditions. Using $LiMn_2O_4$, $LiNiCoO_2$, $LiFePO_4$, or other new cathode materials instead of $LiCoO_2$ might solve the safety issue, and their cost would be less, too.

China's R&D on new Li-ion battery materials has been making good progress. The high-power pack of Li-ion batteries has been used for EVs. For example, in 2007, Tongshen New Energy Company in Jilin Province started a project to develop new types of Li-ion battery materials. It announced that its Li-ion battery made of $LiFePO_4$ composite could be cycled 1,000 times and still keep more than 90% of its capacity. The energy density of the composite material has been higher than 140 mAh/g. It passed safety tests for pressure, short-circuiting, overheating, and overcharging.

2.7.1. Separator Development

The development of separators is important to the development of Li-ion batteries. Li-ion batteries and cells are smaller or larger, depending on the application. For electronic products like cell phones and digital cameras, batteries need to be very small. To keep high capacity in a small area, the separator should be thinner than 25 μm. A separator's properties have an impact on conductivity, which could then influence the battery's capacity, cycle life, and safety. Polyethylene, polypropylene, and other nonpolar materials have been used as separators, but with low conductivity and low absorption of electrolytes. Hence, one research direction is to modify the surface of those separator materials with some plasma or ultraviolet treatment to improve some of the properties.

For E-bikes, EVs, and power tools, high-power and high-capacity batteries are needed. One battery pack may be composed of tens or hundreds of cells. To address safety concerns, 40-μm-thick separators are currently used. Degussa Company of Germany is producing a separator that consists of an organic fiber coated with an inorganic oxide film. This composite separator would greatly reduce safety concerns; that is, even if the organic film melted at a high temperature, the inorganic oxide film would still be functional to protect the battery from short- circuiting.

2.7.2. Progress of Li-Ion Battery Technology for Electric Vehicles

A secondary battery with a thermal management system is one of the key technologies in EVs. In past years, secondary lead acid, Ni-Cd, and Ni-MH batteries did not deliver a range or cycle life that would enable the commercialization of EVs. Li-ion batteries could be a core technology for EV applications, since they are lightweight and have a high energy density, high power, long life, low self-discharging rate, and wide temperature range. The disadvantages of Li-ion batteries are their high cost and safety concerns, but new materials — like $LiMn_2O_4$, $LiFePO_4$, and lithium vanadium phosphate ($LiVPO_4$) — might solve those issues.

Parameters that affect the performance of Li-ion batteries are battery life, use at low temperatures, overheating, overcharging, their relatively high cost, and the difficulty of diagnosing problems. The long-term focus is to develop both materials and systems. On one hand, the reliability of laboratory data on parameters like acceleration, charging time, and driving range issued by companies needs further review. Measurements need to be made under complex practical driving conditions. When EVs are mass-produced, quality control (QC) must be improved. On the other hand, China still needs to import separator materials, which accounts for 30% of the battery cost.

It is believed that in the 1990s, EV performance satisfied the needs of average consumers, even though the energy density of the battery then was lower than it is now. The main reason that EVs were not commercialized was the short life of the secondary batteries. Now, the acceleration and driving range of EVs are reaching the levels achieved by conventional vehicles. The high cost and relatively low life of secondary batteries are still the issues to be overcome before EVs are commercialized. To further improve energy density and safety and reduce cost, more R&D on new electrode materials and manufacturing processes for lighter and thinner batteries are needed.

2.8. Li-Ion Battery Industry and Raw Materials

Since 2001, the production of Li-ion batteries has increased by 40% annually. The Irish Research and Markets Company predicts that the volume of Li-ion batteries produced in China (including districts of the Chinese mainland, Hong Kong, and Taiwan) is larger there than in anywhere else in the world. With regard to the global production of Li-ion batteries, the Chinese mainland accounts for 16.9%, and Taiwan accounts for 6.9%. A census conducted by the China Chemical and Physical Power Sources Industrial Association found that the Chinese mainland produced 950 million Li-ion secondary battery cells in 2006 and about 1 billion in 2007. In 2006, 2 billion Li-ion battery cells were produced worldwide. It is projected that in 2010, 2.66 billion cells will be produced. As the cost of the battery has gone down and its performance has improved, many automobile companies have been investing in and begun integrating them in vehicles. Table 2-3 lists companies that conduct R&D on and produce Li-ion batteries.

More than 20 large automobile companies throughout the world are investing in R&D on Li-ion batteries. For instance, FUJI Heavy Industries, Ltd., and NEC are cooperating to explore cost-effective and safe Mn-based Li-ion batteries. This battery technology could have a life of 12 years and reach 100,000 km under an EV environment. Panasonic is exploring a small, high- capacity, fast-chargeable Li-ion secondary battery. Nanotechnology is used to make the anode material, which could uniformly absorb the Li ion. This would allow the battery to be charged to 80% of its capacity in 1 minute and to 100% in 6 minutes. Johnson Controls established an R&D Division for EV Li-ion batteries at its Milwaukee, Wisconsin, location in September 2005. In 2006, Johnson Controls invested 50% to co-found Johnson Controls – Saft Advanced Power Solution (JCS). In 2006, JCS got a 2-year contract from DOE to work with USABC on EV Li- ion batteries. JCS has contracts to supply high-power Li-ion batteries to automobile companies as well.

China has several criteria for conducting R&D on Li-ion batteries and improving 2010 targets set by USABC. The country has developed safe and lighter (5–10-kg) Li-ion batteries for electric bikes and electric motorcycles and 50–80-kg batteries for EVs, HEVs, and PHEVs with a driving range of 80 km. Li-ion and polymer Li-ion batteries have a large potential market because of they have a higher energy density, longer cycle life, lighter weight, smaller volume, and better safety than traditional batteries.

The factors driving the development of Li-ion batteries in China are as follows:

- The demand for notebooks and cell phones is steadily increasing, as is the demand for Li-ion batteries in the power tool, digital camera, toy, mobile DVD, and other sectors.
- The application of Li-ion batteries in power tools is occurring faster than predicted. The competition among Sony, Sanyo, Samsung, and E-One for LiMn2O4 batteries and their competition with the A123 Company for LiFePO4 batteries increases their spread.
- The development of Li-ion batteries for powering EVs has been fast. IT researchers in Japan state that the battery technology is getting mature and that it should be in mass production in 2009–2011.
- Besides saving energy, HEVs help save the environment. A 2003 study indicated if oil prices stayed at the current level, 5–6% of cars sold in the United States in 2010 would be hybrid. It is estimated that 80% of the market in 2015 will be hybrid cars.
- EVs need very reliable and low-cost batteries. In China in the past, the small production scale, high cost, poor reliability, and low applicability to cars of EV batteries hindered their mass production and development. Now, China's huge E-bike market offers an excellent opportunity for the Li-ion battery industry to expand. Annual sales of 20 million E-bikes will promote the EV industry.

Table 2-3. Li-Ion Battery Company Products.

Company	Summary of Products
Degussa AG/Enax	In June 2005, German Degussa AG and Japanese Enax invested 50% each to establish the Degussa Enax (Anqiu) Power Lion Technology Company in China. It produces Li-ion battery electrodes and EV Li-ion battery electrodes for China, Europe, Japan, and other countries.
Johnson Controls – Saft Advanced Power Solution (JCS)	JCS is the joint venture that resulted from the merger of Johnson Controls and Saft in January 2006. Saft started research on EV Li-ion batteries in 1995. Johnson Controls began producing EV Li-ion batteries for testing in Milwaukee, Wisconsin, in 2005.
NEC Lamilion Energy	In March 2006, NEC supplied an EV manganese (Mn) -based Li-ion battery. Its life was 800 cycles with 2,700 W/kg power at 25 Ca and 10 s. Semivolatile organic compound [SVOC] emissions were 50% higher than the standard based on tests at an equivalent distance of 150,000 km for 10 years.
Sanyo Electric	In March 2006, Sanyo's Japanese factory supplied 1,000 packs of Li-ion batteries for EV testing. They were to be mass-produced in 2007.
Panasonic EV Energy	In October, 2005, Toyota increased its investment in Panasonic EV Energy from 40% to 60%. It was to install plug-in Li-ion batteries in its Prius model in 2008. Mass production and practical performance are being reviewed.
GS Yuassa	In March 2004, GS Yuassa started selling two Li-ion battery models — E-on EX25A (cell) and EX25A-7 — for EVs and for uninterruptible power supplies (UPSs).
Hitachi Vehicle Energy, Ltd.	Hitachi Vehicle Energy was established in June 2004. It produces Mn-based Li-ion batteries for EVs. In June 2006, it developed a small (10% less space), low-cost (12.5% less expensive) Li-ion battery model (48 cells).
Litcel	In 2006, Litcel developed an Li-ion B4-40 (pack) for EVs and installed it in a Mitsubishi Colt-EV for testing. The driving range was 150 km per charge; the target is 240 km in 2010.

[a] C is the hourly capacity rating where 1 C = 1 hour, measured in ampere-hours (A·h).

In recent years, BYD, Lishen, and other Li-ion battery companies have been growing very rapidly. In 2006, exports of Li-ion secondary batteries, which amounted to more than 1 billion cells, were valued at more than U.S. $2.98 billion. This represented an annual increase of 34% and 29% for BYD and Lishen, respectively. Sony exports U.S. $256 million worth of these batteries. BYD and Lishen are the only Chinese-owned companies out of the top-10 exporting companies in China. Japanese, Korean, and Taiwanese ventures are still the main exporting forces for Li-ion batteries, accounting for 60% of the exports. In 2007, these exports were expected to increase by 25%.

In the Li-ion battery market, the competition for cell phones and notebooks has been intense. The price of Li-ion batteries is going down, and the room for profits has greatly shrunk. The new national standard would eliminate small-scale production lines and promote companies with scale and technology advantages. If, however, a "battery explosion" was to occur, it would force consumers to pay attention to the safety of the batteries, and the consumers would rely more on brands with a good reputation.

Because of technology and cost demands, Li-ion batteries used to be small. As rapid progress in the development of key materials and technologies occurs, larger batteries and battery packs will be applied in transportation vehicles, power tools, toys, lights, electric boats, and other products. The Li-ion battery industry is becoming an important sector. The key issue limiting the battery's application is its safety. Its safety relies on breakthroughs related to materials, as follows:

- The selection of safe cathode materials (LiMn2O4 could a choice),
- The selection of separators,
- The protection of valves from explosion, and
- Protective circuits.

2.9. Battery Standards

The Li-ion battery industry needs standards that balance product safety and performance. So far, various standards from China and abroad have been used for testing. Standards have covered short circuits, overcharging, overdischarging, vibration, punching, pressing, dropping, the heating box, a low-pressure atmosphere, temperature cycles, and other parameters that simulate normal and abnormal situations for battery applications. The objectives of the tests are to have good operating criteria and achieve ease of operation. The standards also provide a path for designing battery technology that performs safely and acceptably. The safety standards in China still need to be developed at an international level.

2.9.1. Status of Chinese Li-Ion Battery Standards

Here are the eight current safety standards for Li-ion batteries in China:

1. GB/T18287-2000 is for cell phone Li-ion batteries.
2. GB/T19521.11-2005 is for checking Li-ion battery packs.
3. GB/Z18333.1-2001 is for EV Li-ion secondary batteries.
4. YD1268.1-2003 gives safety requirements and test methods for mobile phone Li batteries.
5. SJ/T11169-1998 is for Li batteries.
6. QB/T2502-2000 is for Li-ion secondary batteries.
7. QC/T743-2006 is for EV Li-ion secondary batteries.
8. SN/T1414.3-2004 is a safety testing method for exporting and importing secondary batteries. Section 3 covers Li-ion secondary batteries.

The first three are national standards. The last five are from industrial sectors: the post office, electronics, light industry, automotive, and product QC sectors.

Three standards for Li-ion battery safety are still to be issued:

1. GB/Txxx-200x/IEC62133:2002 will be for alkali or other nonacid electrolyte- based secondary batteries and battery packs (portable sealed secondary batteries and battery packs).
2. GB/T18287-200x will be for mobile phone Li-ion secondary batteries and battery packs.
3. SJ/Txxx-200x will be for notebook mobile power source systems.

The first two are national standards; the third one is for the electronic industry.

2.9.2. Status of International Li-Ion Battery Standards

The influential international standards are International Electrotechnical Commission (IEC) standards; Underwriters Laboratories, Inc. (UL) standards; and Institute of Electrical and Electronics Engineers, Inc. (IEEE) standards. The six current international standards related to Li-ion secondary batteries are as follows:

1. IEC62133:2002 is for alkali or other nonacid electrolyte-based secondary batteries and battery packs (portable sealed secondary batteries and battery packs).
2. IEC62281 :2004 is for safely transporting lithium primary and secondary batteries and battery packs.
3. UL1642:2006 is for lithium batteries.
4. IEEE1625:2004 is for notebook secondary batteries.
5. IEEE1725:2006 is for cell phone secondary batteries.
6. IEEE1 825:2006 is for secondary batteries in digital cameras and video cameras.

2.10. Market Analysis of Raw Materials for Li-Ion Batteries

2.10.1. Market for Cathode Materials

As the range of applications for Li-ion batteries gets bigger and high voltage and high energy density are required, the safety and high capacity of Li-ion batteries will also get attention. LiCoO2 is still the most attractive cathode material for Li-ion batteries, but safer cathode materials with higher capacities and lower costs are being explored and developed. Cathode materials for smaller Li-ion batteries are mainly LiCoO2 and $Li_{Ni_xCo_{1-x}O_2}$, while cathode materials for larger Li-ion batteries are spinel LiMn2O4 and layered LiMnO2. It is estimated that the consumption of cathode materials all over the world is 8,000–10,000 tons. China's supply of this material is quite adequate. MGL, YuYao JingHe, Shanshan Tech, Guotai Huarong, and other companies compete in supplying cathode and anode materials and electrolytes. Downstream companies, such as Shenzhen BYD, BAK, B&K, and TCL Jingneng, have a 20% share of the global Li-ion battery market.

Since the intercalation compound formed by TiS2 and Li was discovered in the 1 970s, Sony has been using $LiCoO_2$ as cathode material and graphite as anode material. Negative selenium (Sn)- and silicon (Si)-based materials are still being evaluated in laboratories, while the cathode materials $LiCoO_2$, $LiMn_2O_4$, $LiNi_xCo_{1-x}O_2$, $LiMn_yNi_xCo_{1-x}O_2$, and $LiFePO_4$ are currently being used in batteries that are under development.

$LiCoO_2$ is still the most mature and commercialized cathode material. It has some advantages. It can be easily processed. It has high power density and relatively high energy density with a stable structure. It has good cyclability (charging and discharging ability) and high-voltage output. Although this material will be difficult to replace, it may nevertheless be replaced in 5–10 years because of its high cost, low availability, and poor safety. In power batteries, it would be replaced by $LiMn_2O_4$ and $LiFePO_4$. In communication applications, it would be replaced by $LiNi_xCo_{1-x}O_2$ and $LiMn_yNi_xCo_{1-x}-yO_2$.

$LiFePO_4$ has been under development in recent years. Its low cost, high level of safety, good structural stability, and good cyclability make it very attractive for power batteries and energy storage. Its vibrational density, volume energy density, conductivity, low-temperature discharging, and high-rate discharging need to be improved, however. The capacity of $LiNi_xCo_{1-x-y}O_2$ is more than 30% higher than that of $LiCoO_2$. It costs less, but it is relatively difficult to synthesize and has a low density, low charge-discharge efficiency, and poor safety. $LiMnyNi_xCo_{1-x-y}O_2$ is a high-capacity cathode material (>180 mAh/g) with good safety, a relatively low cost, compatibility with the electrolyte, and good cyclability. It would most likely be applied in communication and small power batteries and perhaps even in large power batteries.

2.10.2. Market for Anode Materials

Carbon materials are still playing an important role in the negative materials used in Li-ion batteries, but market competition is pushing anode materials toward having a higher capacity and lower cost. Nano-sized carbon material is a potential anode material. Carbon microsphere material is produced in China as a cell phone anode for Li-ion batteries. The carbon materials produced by Shanshan Tech and Tianjin Tiecheng have a capacity of 300–400 mAh/g, and they cost about two-thirds as much as the imported products. In the power battery field, the life of the carbon anode material could be greatly shortened because of the large volume change that results from a large current. The higher capacities of anode materials like Sn and Si metal or oxide materials have been studied. But Sn and Si materials expand by as much as 300–400% during the charge-discharge process, which causes poor cyclability. The use of nano-alloyized Sn-based materials might resolve this issue; this has been studied by Sony and in the Chinese 863 Plan. The three-dimensional current collector could be another target of anode materials research; it could reduce inner resistance and polarization and lessen the "volume effect" to improve cyclability and charge-discharge efficiency.

2.10.3. Market for Separators

A Li-ion secondary battery is usually composed of a cathode, an anode, an electrolyte, a separator, and a case. The separator is an important material in Li-ion batteries; it transports ions and protects cathodes and anodes from short-circuiting. Its performance affects the interface structure and also inner resistance, which influences the battery's capacity, cyclability, and safety. There are a few kinds of separators: a one-layer film; a two layer film of polypropane (PP) and polyethylene (PE); and a multilayered film that is a composite of PP and PE, such as PP/PE/PP. Only a few countries, like Japan and the United States, have the technology for mass-producing separators. R&D on separators began much later in China, and the country's supply of separators still relies on imports, which results in high costs. The average price for a separator is 8–15 ¥/m^2, which accounts for one-third of the battery cost. On the basis of annual sales of 1 billion Li-ion battery cells, the consumption of separators is 300– 500 million m^2, with a market value of 1–1.5 billion ¥.

The current processes for producing PE are classified as wet and dry methods. In the dry method, the process is classified further as either a single-direction or double-direction stretch. The dry method with single-direction stretch is described in U.S. Patent 3426754 (1970), owned by USA Cellanse Company. After a few decades of development, this process is quite mature in the United States and Japan. USA Celgard Company and Japanese UBE, which produce single-layer PP and PE and triple-layer PP/PE/PP, are developing manufacturing processes for these materials. Industrialization of the single-direction stretch process in China is very slow because it is limited by foreign patents. In 2004, China began to own some patents on the process, in which some additives were added to improve the process. In Hangzhou, a production line was installed to produce separators by using the modified process.

The wet method is also called the phase separation method or heat-induced phase separation method. In this method, polyolefin, after the addition of small molecules with a high boiling point, is heated until it has melted uniformly, cooled down to the phase separation stage, stretched, and then extracted out of the small molecules with an organic agent in order to obtain the separator material with micropores. Asahi Kasei Corporation, Tonen Chemical Corporation, and USA Entek use this method to produce single-layer PE. In 2004, Foshan Plastics started a production line by using this method.

The dry method with double-direction stretch (CN1062357) was explored by the Chemistry Institute of the China Science Academy, which has been producing single-layer PP. In 2005, the Institute cooperated with New Time Science & Technology Company to start a production line (6 million m^2 of PP); this was done in November 2007. In Xinxiang, Henan Province, there is another production line (15 million m^2) that is using this method; it started to supply PP in 2007.

In summary, the separator market mainly consists of foreign brands, although some Chinese brands have emerged. Since Chinese separators entered the market, their price per square meter has decreased from about 15–20 ¥ in 2003 to about 8–15 ¥.

2.10.4. Market for Electrolytes

Statistics indicate that the production of electrolytes for Li-ion batteries increased from 1,920 tons in 2004 to 7,142 tons in 2007, representing an annual growth rate of more than 90%. In the coming years, the demand for electrolytes will continue to grow rapidly as global production of Li-ion batteries moves to China and the demand for power batteries increases.

Electrolyte production is developing very fast in China since its start at zero, and it is playing an important role in the development of the Li-ion battery industry. Electrolytes are one of the four key materials (cathode, anode, separator, electrolyte) in a Li-ion battery, which is composed of high-purity solvent, such as lithium hexofluorophosphate ($LiPF_6$), and additives.

2.10.4.1. High-Purity Solvent

The organic solvent for the electrolyte in a Li-ion battery must be a non-proton solvent. To keep the battery system safely working under a wide range of temperatures, the solvent should have a low melting point, high boiling point, and low vapor pressure. A multi-chemical solvent is needed to satisfy these requirements. In general, common electrolytes are composed of high-dielectric-coefficient carbonate solvents and low-viscosity, chainlike carbonate solvents. Low-temperature electrolytes are composed of low-melting-point solvents and high-dielectriccoefficient, cyclo-like carbonate solvents. High-power battery electrolytes use high-boiling-point solvents; high-flashpoint solvents; and high-dielectric-coefficient, cyclo-like, carbonate solvents.

The organic solvent for electrolytes must have high purity and be very low in moisture content. The use of a technology that removes impurities and moisture is a key step in making electrolytes. Suitable solvents include DMC (dimethyl carbonate), DEC (diethyl carbonate), PC (propane carbonate), EC (ethylene carbonate), and EMC (ethyl methyl carbonate).

Before 2000, all solvents were imported by China. Now, companies in China can supply >99.95% high-purity solvents and satisfy all the country's needs. The main manufacturers are Liao Yang Kong Lung Chemical Industry Ltd.; Hebei Tanshan Zhaoyang Chemicals; Shandong Taifeng Mining Group Co., Ltd.; Shandong Shida Shenghua Chemical Co., Ltd.; Jiangsu Taipeng Medicine and Chemical; The Reagent and Chemical Plant of Taixing City; and others.

2.10.4.2. LiPF$_6$ Electrolytes

LiPF$_6$ is a good solute for Li-ion battery electrolytes. It has to have high purity and low moisture. Manufacturing it is difficult, highly dangerous, and highly technical because of its strong moisture absorption. The market for LiPF6 was dominated by Japanese companies. In 1997, Tianjing Chemical Engineering Research Institute finished pilot tests for manufacturing LiPF$_6$, and in one year, 2000, it installed 4 tons of demonstration equipment. In 2002, the Tianjing Institute and Xingtai Mining Group co-founded Tianjing Jinniu Energy Materials Company. In 2003, this company installed 80 tons of equipment for manufacturing LiPF$_6$.

2.10.4.3. Additives

Although the amount of additives that is added to electrolytes is small, the additives play an important role in improving the capacity, cyclability, swelling resistance, flame resistance, and safety of Li-ion batteries. Vinylene carbonate (VC) is an important additive that is now being explored. It is used to reduce ethylene and propene gases and improve solid electrolyte interphase film and safety. Flame-resistant additives are also important in preventing organic electrolytes from burning at high temperatures. P-containing and B-containing additives with a high boiling point and flashpoint could be used. Some additives are used to prevent overcharging. Before 2004, all additives were imported. The price of a kilogram of VC was more than 6,000 ¥. Most of the additives can now be produced in China.

2.10.4.4. Electrolytes

The five requirements for electrolytes in Li-ion batteries are as follows:

1. High ion conductivity of 10^{-3} to ~2×10^{-3} s/cm (Li-ion transfer coefficient is close to 1);
2. Large stable electrochemical potential, with an electrochemical window of 0 to ~5 V;
3. Wide operating temperature range;
4. Good chemical stability (do not react with current collector and active materials); and
5. Not very poisonous and bio-decomposable.

Depending on the solvents, solute amounts, and additives used, there are different brands of electrolytes and different electrolyte products. Electrolyte and battery manufacturers have been exploring new electrolytes by synthesizing new solutes, synthesizing higher-dielectric-coefficient solvents, and making new additives. The companies that produce electrolytes include Zhangjiakong Guotai Ronghua; Tianjin Jinniu Energy Materials; Shantou Goodsun; Guangzhou Tinci Silicone Technology Co., Ltd.; and Beijing Chuangya.

The production capacity of Li-ion batteries is estimated to have reached 1 billion cells in 2008. The demand for LiPF6 was between 400 and 500 tons. More than 3,000 tons of electrolytes were needed.

2.11. Electric Vehicle and Electric Bicycle Market

As hybrid automobile technology is maturing, the power battery market for EVs is expanding. There are four "bike cities" in China: Beijing, Tianjin, Shanghai, and Chengdu. There are 10.5 million bikes in Beijing, 9.7 million in Tianjin, 9.2 million in Shanghai, and 7.5 million in Chengdu.

A survey conducted by Tongji University in Shanghai indicates that as many as 76% of Chinese citizens in big cities would like to use E-bikes instead of regular bikes, which means 350 million of China's 450 million bike customers would like to be E-bike customers. More than 2,000 companies produce E-bikes. In 2001, only seven of them produced more than 20,000 E-bikes per year. In 2001, production amounted to 444,800 E-bikes. In 2002, it was 802,200. In 2006, it was 19.2 million. Most of the companies produce fewer than 10,000 E-bikes per year. The E-bike companies with production capacities of more than 500,000 E-bikes per year are Tianjin Fushida, Taiwan Lujia, Zejiang Xiaofeige, and Haian Xindazhou. Those producing 300,000 E-bikes per year are Wuxi Yilin, Shanhai Yongjiu, Tianjin Dushifeng, Tianjin Daanmonaduo, Tianjin Damin, and Hongkong. Daben, Shanhai Saifeng, and Sichuan Fushi produce 200,000 E-bikes per year. In 2007, the rising prices of lead and China's national policy sharply cut the E-bike market. However, in the long term, the E-bike market will be very significant. The potential is for 400 million consumers and for 1.3 trillion ¥ in revenue for E-bikes in China.

The growing demand that is expected for batteries would be for Li-ion batteries. In 2003, E-bike production was 4 million; in 2004, it was 7 million; in 2005, it was 10 million; and in 2006, it was 19.2 million and involved more than 2,000 companies. Figure 2-1 shows the history of and projections on E-bike production in China. The growth of E-bikes would drive the demand for Li-ion batteries. So far, batteries for E-bikes have been lead acid, Ni-Cd, Ni-MH, and Li-ion batteries. The Li-ion battery has advantages over the others in every aspect but cost. As the scale of production of Li-ion batteries and the number of applications for them increase, their cost will go down, and they will replace lead acid and other battery technologies to power E-bikes.

Tianjin, Nanjin, Chengdu, Shenyang, Zhengzhou, Shenzhen, Hangzhou, Suzhou, Jinan, Taiyuan, and other big cities have opened up their roadways to E-bikes. In some cities like Beijing, Guangzhou, Changchun, Fuzhou, and Hehui, they were once was banned but are accepted today.

There are four kinds of power batteries for e-bikes: lead-acid, Ni-MH, Ni-Cd, and Li-ion power batteries. Even though there has been technical progress in developing lead acid, Ni-Cd, and Ni-MH batteries, they lack market compatibility and consumer acceptance. Thus, Li-ion batteries have a very high potential for being the power battery. E-bike production has reached 400,000 per year. China consumers own 400 million bikes, of which 20% are E-bikes; there are 80 million E-bikes. In the coming two to three decades, E-bikes will replace bikes, and Li-ion batteries will power those E-bikes.

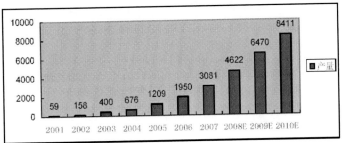

Purple bars indicate production. E means estimated. Units are in thousands. Data source: Chinese National Bureau of the Census.

Figure 2-1. E-Bike Production in Recent and Upcoming Years in China.

2.12. Supply of Raw Materials

The main raw material used to produce Li_2CO_3 is salt lake brine. Li_2CO_3 producers must get the right for resource exploration from the Chinese government. Most salt lakes contain a large amount of magnesium (Mg) and a small amount of lithium (Li), which makes separation and extraction difficult. The main market was three Li_2CO_3 manufacturers: SQM S.A. (formerly Sociedad Quimica y Minera de Chile, then SQM Chemicals) and the U.S. companies FMC (formerly LCA) and Chemetall GmbH (formerly Foote Minerals and then Cyprus Foote) before a few large Chinese production projects began. SQM's production capacity is 28,000 tons/yr; FMC's is 22,000 tons/yr; and Chemtall's is 18,000 tons/yr. These three companies account for 73% of global production capacity. SQM added 12,000 tons/yr in 2008; FMC added 5,000 tons/yr in 2009; and Chemtall added 5,000 tons/yr in 2008.

In China, Li_2CO_3 is being produced from salt lake brine in two large projects. MGL is planning to produce 35,000 tons/yr; 25,000 tons/yr has been produced since 2007. Tibet Mining Group is planning to produce 30,000 tons/yr; a 5,000-ton/yr production line was established in 2005.

3. LITHIUM-ION BATTERY TECHNOLOGY FOR TWO-WHEELERS IN CHINA

3.1. Overview

In 2006, 20 million electric bikes were made in China. Currently, there are 50 million battery-operated bicycles on the road in China. Of these, only a very small percentage operate on Li-ion batteries; the rest use lead acid batteries. About 2,500 companies in China produce electric two- or three-wheelers. All of the large EV manufacturing companies have E-bike models with Li-ion batteries, but their performance-to-price ratio is still not compatible with that of E-bikes with lead acid batteries. This is the key reason that the Li-ion battery bikes are not in mass production yet. The price of Li-ion battery packs is three to four times higher than that of lead acid batteries. The increase in the price of a ton of lead from 15,000 ¥ in 2006 to

25,000–34,000 ¥ in 2007 may enhance the market for two- or three-wheelers with Li-ion battery packs.

In Shenzhen, more than 150 companies make secondary batteries, including Li-ion, Ni-MH, lead acid, and Ni-Cd batteries. Of these companies, 95% are privately owned, and half of them work on Li-ion batteries. In 2006, battery production had increased from 2003 levels; the increase for Li-ion batteries was 152.6%; for Ni-MH batteries, 117.2%; for lead acid batteries, 22.6%; and for Ni-Cd batteries, 4.7%. The revenue from Li-ion batteries was 58.1% of total revenues from secondary batteries in Shenzhen. Li-ion batteries are a priority product supported by the Shenzhen government. In 2006, 10 Shenzhen battery companies were listed in the top 100 Chinese battery companies. Their total revenues were more than 950 million ¥, accounting for 73.6% of the Shenzhen secondary battery industry. They were (1) BYD; (2) B&K; (3) BAK; (4) Shenzhen Highpower Technology Co., Ltd.; (5) Shenzhen Central Power Technology Co., Ltd. (mainly lead acid batteries); (6) Shenzhen EPT Battery Co., Ltd.; (7) Shenzhen Sunnyway Battery Technology Co., Ltd. (mainly lead acid batteries); (8) Shenzhen HYB Battery Co., Ltd.; (9) Lexel Battery (Shenzhen) Co., Ltd. (mainly lead acid and Ni-MH batteries); and (1) Shenzhen Xwoda Group Co., Ltd.

3.2. Institutions and Companies Visited

3.2.1. Beijing Institute of Technology

The Beijing Institute of Technology (BIT), EV Center of Engineering and Technology, is located at No. 5 South Zhonggancun South Street, Haidian District, Beijing 100081. BIT cooperated with MGL and Tsinghua University to make electric buses for the Beijing Olympics in 2008. BIT is the most prestigious institute in China. It has excellent laboratories, with the most modern equipment for testing and evaluating each of the components of an electric bus. BIT can also evaluate electric buses and passenger cars on a dynamometer and on controlled tracks.

A typical Li-ion battery for a large-bus application consists of 108 cells in series and of four banks in parallel to provide 400-A·h capacity with a nominal 388 V. These batteries charge to 4.2 V and discharge to less than 3 V.

BIT is working on manufacturing electric buses with the following companies:

- Beijing Beifang Huade Niopolan Bus Company, Ltd.;
- Jinghua Bus Company, Ltd.; and
- BIT Clean Electric Vehicle Company, Ltd. (a separate company within BIT).

The technology for the electric buses is described here for buses having a rare-earth, charging-flux, permanent-magnet DC motor:

- The rare-earth permanent-magnet material with magnetic winding combined with excitation can make a permanent DC motor.
- The permanent-magnet DC motor will have a rotor that adopts to a no-groove structure.

- The increasing magnetic winding will link to a re-flowing current loop in order to auto-decrease the magnetic field.

The controller for the rare-earth, charging-flux, permanent-magnet DC motor:

- Is integrated with a high-frequency, power-winding motor control;
- Auto-decreases the frequency modulation;
- Increases the current of the closed-loop controller; and
- Recovers regenerative braking energy.

System parameters are as follows:

- Efficiency of more than 92%, with 80% of the area having high efficiency for 84.4% of the operating time;
- 75-kW steady-state/125-kW peak maximum moment of system at 1,200 nm; and
- Line control with two-speed gear box.

Specifications for the electric ultra-low-floor bus are as follows:

- Li-ion battery: 388.8 V at 400 A·h;
- Power driving system: three-phase, asynchronous, alternating-current (AC), 100-kW motor;
- Wheel base: 5,800 mm;
- Wheel span: 2,340 mm for front wheels, 3,440 mm for rear wheels;
- Curb mass/full-load mass: 12,930/16,000 kg;
- Maximum velocity: 91 km/h;
- Driving range at 40 km/h: 210 km;
- Acceleration time from 0 to 50 km/h: 20.7 s; and
- Braking distance at 30 km/h: 8.2 m.

BIT evaluated 12 electric buses for use in the 2008 Beijing Olympics. Nearly all of the laboratory equipment for evaluating electric buses was imported from Germany, Japan, and the United States. BIT is 100% supported by the Chinese federal and state governments. It holds several patents for electric drivetrains for buses and for passenger vehicles.

3.2.2. CITIC Guoan Mengguli Corp

CITIC Guoan Mengguli Corp. (MGL) is located in Beijing Zhongguancun Science Park. Beginning in 2007, MGL (or Beijing MGL) focused on delivering exclusively Li-ion power batteries to 50 electric buses for the 2008 Olympics. The 50 buses, on three inner circular lines, were to provide transportation between the Olympic Village, Media Village, and some stadiums. Performance data on MGL's EVs with a 100-A·h, Li-Mn power battery are as follows:

- Motor power: 30 kW;
- Maximum speed: 120 km/h;

- Acceleration time from 0 to 50 km/h: 7 s;
- Driving range per charge: 264 km;
- Charging time: 2–4 h; and
- Energy consumption: 18 kWh per 100 km.

In the spring of 2004, MGL had begun delivering 400-A•h, 400-V LiMn2O4 batteries for testing for operating electric buses. It was expected that by the end of 2008, MGL would have made 500 electric cars to be used as taxis by a local city government.

MGL is engaged in R&D on and production of new composite metal oxide materials and high-energy-density Li-ion secondary batteries. The primary investor in MGL is CITIC Guoan Group, a wholly owned subsidiary of China Zhongxin Group (CITIC). The CITIC Guoan Group has operations in various industries, including IT, new materials, mineral resource surveying, tourism, and real estate. Ratified by Deng Xiaoping, CITIC was founded in October 1979 by Rong Yiren, former Vice Chairman of the PRC. Having experienced growth for more than 20 years, CITIC is now a large-scale international enterprise group with total assets of 700 billion ¥.

MGL is China's largest manufacturer of the Li-ion cathode material $LiCoO_2$ and is first in line to market the new cathode materials $LiMn_2O_4$ and $LiCo_{0.2}N_{i0.8}O_2$. Being quality-oriented, MGL has been certified to both the New and Hi-Tech Enterprise standards and ISO 9001:2000. MGL's unique synthesis method simply and efficiently produces cathode materials of superior electrochemical performance and reliability in an environmentally friendly way. Since incorporation, MGL has smashed the monopoly of China's Li-ion battery cathode materials market held by foreign manufacturers, and it now stands at the forefront of that industry. Besides cathode materials, MGL also produces high-capacity, high-energy-density Li-ion secondary batteries for power and energy storage, with capacities ranging from several A•h to several hundred A•h. As China's first and only power battery manufacturer, MGL is now setting the global pace by presenting high-capacity Li-ion secondary batteries, which have been successfully applied to Beijing's trial fleet of electric buses.

To ensure sustainable and steady development, MGL has built up a modern R&D department in Beijing. Through the combined efforts of MGL's staff members, MGL is able to contribute more and more to social progress and development.

3.2.3. Tsinghua University

The Department of Automotive Engineering of Tsinghua University is located in Beijing. Tsinghua University cooperated with MGL, BIT, and other companies to make electric buses for the 2008 Beijing Olympics. The specifications for the battery packs for these buses were as follows: 80 A•h, 30 kWh, and 300–400 V for Ni-MH battery packs and 100 A•h, 30–40 kWh, and 300–400 V for Li-ion battery packs.

The Department of Automotive Engineering started doing research on EVs in 1995, on HEVs in 1998, and on fuel cell electric vehicles (FCEVs) in 1999. It is working with five battery companies on Li-ion batteries for vehicular applications: B&K Battery Company; Thunder Sky; MGL New Energy Technology Co., Ltd.; Oriental Polymer; and Huanyu Battery Co. The department has an excellent facility to test and evaluate complete vehicular systems as well as batteries at the module and pack level. Battery testing and evaluations are conducted by using the following: the national standards of the PRC; the USABC battery testing manual; the Partnership for a New Generation of Vehicles (PNGV) battery testing manual; the FreedomCAR battery testing manual; the testing standards of Japan; and other testing standards, such as those developed by the American Society of Mechanical Engineers (ASME).

Currently, the collaborators are evaluating hybrid and fuel cell hybrid buses with Li-ion batteries. The batteries are 100 A•h, and a pack contains 30 cells. This evaluation is being conducted for the 863 Program. Oriental Polymer in Beijing is supplying an 1 8-kW fuel cell, and Thunder Sky is providing a 100-A•h Li-ion cell with 100 cells per pack in series. A hybrid bus has been evaluated for 1,250 km without any degradation in battery performance. Similarly, a fuel cell hybrid bus has been evaluated for 2,340 km without any degradation in performance.

Other EV research includes:

- Structural design,
- Parameter matching and optimization of powertrain system,
- Optimization of energy management strategy,
- Controller design,
- Communication network,
- Failure diagnostics,
- Other subsystem testing, and
- EV assembly and road testing.

The department's focus is research on EVs, HEVs, PHEVs, and FCEVs, with an emphasis on battery applications. Two professors, four associate professors, two engineers, five part-time experts, four postdoctoral students, four doctoral candidates, and 13 master's students are involved in the research. The university has six patents and four applications pending in China on battery thermal management, EV controller design, and electronics for vehicles. The department is working with GM on FCEVs, but details on this work were not available.

3.2.4. China Electrotechnical Society

The China Electrotechnical Society is located at 46 Sanlihe Road, Beijing 100823. It has 123,000 members and is a clearing house for electrotechnical research. It conducts studies on battery technology markets for EVs and HEVs.

An investigation by the Electric Vehicle Institution in the Chinese Electrotechnical Society showed that in 2006, 20 million EVs had been made. Currently, China has 50 million battery-operated bicycles on the road. Only a very small percentage of them operate on Li-ion batteries; the rest use lead acid batteries. All of the large EV companies have E-bike models with Li-ion batteries, but the performance-to-price ratio is still not compatible with that of E-bikes with lead acid batteries. This is the key reason that bikes with Li-ion batteries are not yet in mass production. The price of a Li-ion battery pack is four to five times more than that of lead acid batteries.

The capacity of E-bikes with Li-ion batteries is 5–10 A•h. The range is 20–30 km for 24-V batteries and 40 km for 36-V batteries. The key issues to be addressed in making Li-ion power batteries are consistency and QC. Recently, the China Electrotechnical Society completed a preliminary study on E-bike technology. The study was conducted for a company in France, EDF. A copy of the report was not available.

3.2.5. China Automotive Technology & Research Center

In 1985, upon the approval of the China National Science and Technology Commission, the China Automotive Technology & Research Center (CATARC) was established to respond to the state's need to manage the auto industry. It is now affiliated with the State-Owned Assets Supervision and Administration Commission (SASAC). It has 1,476 employees, of which 638 are technical professionals, including 54 professor-level senior engineers, 26 doctors, and 207 senior engineers. Assets total 900 million ¥, and CARTARC covers 240,000 m^2 of land.

As a technical administrative body in the auto industry and a technical support organization to governmental authorities, CATARC assists the government in various activities, such as formulating auto standards and technical regulations, product certification testing, quality system certification, industry planning, research on industry policies, information services, and research on common technologies. CATARC has built up the amount and breadth of its competency by setting up testing laboratories and research departments and attracting technical talent.

3.2.6. BYD Battery Co., Ltd.

The Baolong Plant of BYD Battery Co., Ltd. (BYD) is located at No. 1 Baoping Road, Baolong, Longgang, Shenzhen 518116, PRC. BYD's cell development department (Department 1, Li Battery Division, Group 2) produces 120,000 Li-ion battery cells per day. BYD has a concept electric car (60–200 A•h, 380 V) with a charging station. Its 60-A•h, 70-kW, 330-V PHEV — F6DM — went on the market in 2008. F6DM uses an iron battery (LiFePO4) with 18 kWh per 100 km and a total power of 125 kW. F6DM can be charged from a home power source or professional charging station. At home, it takes 9 h to charge. At a station, it takes only 15 min to charge the battery 80%.

BYD is the third largest rechargeable battery manufacturer in the world. It specializes in Ni-MN, Ni-Cd, Li-ion, and lead acid cells and chargers with a wide range of applications for power tools, toys, digital cameras, mobile phones, and cordless phones, among other devices. It aims to offer competitive prices and good quality. Involved in manufacturing, wholesale supplies, and retail sales, its products, in addition to the vehicle batteries, include cordless phone batteries, battery chargers, industrial batteries, mobile phone batteries, and two-way radio battery packs.

As a Chinese private enterprise in Hong Kong, BYD has engaged in two major businesses: IT parts manufacturing and automobile manufacturing. Main IT products include rechargeable batteries (Li-ion, Ni-Cd, Ni-MH), liquid crystal displays (LCDs) and liquid crystal modules (LCMs), plastic housing and tools, keypads, flexible printed circuits, cameras, and vibrators. Auto products consist of high-, medium-, and low-end gasoline cars, ranging from 800 to 2,400 cc. BYD has 130,000 employees.

3.2.7. Shenzhen B&K Technology Co., Ltd.

Shenzhen B&K Technology Co., Ltd. (B&K) is located at Hongfu Industry Park, HuaRong Road, Dalang, Longhua Town Bao'an District, Shenzhen 518109, China. Its web site is http://www.bkbattery.com/cn/about.asp. It makes 10-A•h batteries for electric bikes but does not mass produce them yet. Table 3-1 shows the high-rate performance of its LiFePO4 battery, and Table 3-2 shows its power.

B&K is a player in the ever-expanding rechargeable battery field. Today's batteries play an essential part in business, communications, entertainment, and more; in essence, batteries are an inseparable part of our daily life. With more than 4,000 employees and a strong R&D department, B&K is committed to creating high-quality rechargeable Li-ion and Li-polymer batteries.

B&K was founded in November 1999 and has about 4,000 employees. It focuses on Li-ion battery R&D, production, and sales. B&K has a big influence in the Li-ion battery industry, with two famous brands — B&K and Encel — and three series of products: liquid Li-ion batteries, polymer Li batteries, and LiFePO4 power batteries and cylindrical batteries. Its sales are spread over Europe, the United States, the Middle East, Southeast Asia, and more. Its revenue is more than 500 million ¥.

In 1999, B&K started exploring high-energy Li-ion batteries. It established a modern R&D center with advanced testing and experimental facilities and attracted many battery experts from China and abroad. It developed a series of Li-ion battery products. Its polymer Li-ion batteries have been produced since 2004, and its LiFePO4 power batteries have been produced since 2005. Safety issues have been solved. The company has gained ISO 9000, ISO 14000, U.S. UL, Comformité Européene (CE), and the European Commission's ROHS (Restriction of Use of Hazardous Substances) certification.

B&K has expanded exponentially since 1999, from less than 100 to more than 4,000 employees. It produces 750,000 Li-ion batteries per month, and battery production increases at a 30% growth rate. At the end of 2006, the B&K Industrial Park was started; the park will cover 190,000 m² with 220,000 m² of building area. After the park is completed, the company will produce 1 million of Li-ion batteries per day, and revenue will be more than $500 million.

Table 3-1. High-Rate Performance of B&K LiFePO4 Battery.

Rate (C)[a]	Discharge Capacity (mAh)	Rate (%)
0.2	4027.834	100.0
0.5	3982.027	98.9
1	3954.026	98.2
2	3915.324	97.2
3	3946.809	98.0
4	3907.625	97.0
5	3781.542	93.9
6	3548.394	88.1

[a] C is the hourly capacity rating where 1 C = 1 hour, measured in A•h.

Table 3-2. Specifications for B&K Batteries.

Model No.	Capacity (A·h)	Voltage (V)	Dimensions (mm) Thickness	Width	Length	Impedance (mΩ)	Weight (g)
BK268090	12	3.7	26	80	90	8	470
BK2680130	18	3.7	26	80	130	4	680
BK50120130	60	3.7	50	120	130	3	1,800
BK60130150	80	3.7	60	130	150	2	2,200
BK55145255	100	3.7	55	145	255	1	3,600

B&K is building not only its production base but also as a research institute and an education base. Since 2002, the company has had a post-doctoral station with Tsinghua University and Central-South University.

3.2.8. Shenzhen BAK Battery Co., Ltd.

Shenzhen BAK Battery Co., Ltd. (BAK), is located in BAK Industrial Park, Kuiyong Town, Longgang District, Shenzhen, Guangdong 518119. The telephone number is 86 755 8977 0062, and the web site is http://www.bak.com.cn/. Each day, Shenzhen BAK produces 600,000 cells for cell phones, 150,000 cells (type 18650) for notebooks, and 20,000 cells for polymer Li-ion batteries. Li-ion power batteries for electric bikes are still at the research stage. BAK uses four cells (type 26650) of 2.5 A·h in parallel, then 11 in series, to make 10-A·h, 36-V battery packs. The range of the E-bikes is 45–5 0 km per charge. BAK has patents for protective boards for the Li-ion battery packs. The positive material in the battery is LiFePO4.

China BAK Battery, Inc., is a commercial manufacturer of standard and customized Li-ion rechargeable batteries for use in various portable electronic applications, including cell phones, MP3 players, laptop computers, electric bicycles, digital cameras, video camcorders, and general industrial applications. China BAK is the largest Li-ion replacement battery manufacturer in the PRC and one of the top-three largest manufacturers in the world. It was incorporated in Nevada. In Shenzhen, it operates on a 62-acre site with 1.9 million ft^2 of space for manufacturing. At the BAK Industrial Park location, it produces 1 million cells per day, and its production capacity could reach 1.5 million cells by 2011.

3.2.9. Shenzhen Highpower Technology Co., Ltd.

Shenzhen Highpower Technology Co., Ltd., is located at Luoshan Industrial Zone, Pinghu, Longgang, Shenzhen, Guangdong, China 518111. Its web site is http://www.haopengbattery.com/. The company is licensed by Ovonic to make Ni-MH batteries. In 2007, it made 100 million cells, each one costing $1.20–1.30. It is doing research on an 1 1-A·h, 24-V power battery for E-bikes, with a range of 20–30 km. The price of a lead acid battery is 200– 300 ¥, while a Ni-MH battery pack costs 800–1,000 ¥ and has a life of 1–2 years. In 2007, it made between 3,000 and 4,000 packs for E-bikes. Table 3-3 shows the specifications for the Ni-MH battery for electric bikes.

Table 3-3. Specifications for Shenzhen Highpower Technology Ni-MH Batteries.

Model No.	Size	Capacity (A·h)	Dimensions (mm) Height	Dimensions (mm) Diameter	Maximum Discharging Current (A)	Rapid Charge Current (mA)	Rapid Charge Time (h)	Weight (g)
HFR-60DP7000	D	7	61.5	33.0	21	2,100	4.5	145
HFR-60DP8000	D	8	61.5	33.0	24	2,400	4.5	155
HFR-60DP9000	D	9	61.5	33.0	27	2,700	4.5	165
HFR-90DP12000	F	12	91.0	33.0	36	3,600	4.5	235
HFR-90DP13000	F	13	91.0	33.0	39	3,900	4.5	235

Shenzhen Highpower Technology Co. specializes in research on and manufacturing and marketing of Ni-MH rechargeable batteries. It is located in the town of Pinghu in Shenzhen; its neighbors are Hong Kong and Macao. The company has more than 2,174 employees, including 100 company-trained staff members who ensure QC at each step of the production process.

3.2.9.1. Research and Development

To enhance product quality, reduce costs, and keep pace with technological advances and evolving market trends, Shenzhen Highpower Technology established an advanced R&D center. The center not only focuses on enhancing Ni-MH-based technology by developing new products and improving the performance of current products, but it also develops alternative technologies, such as the line of Li-polymer batteries currently being developed for higher-end, high-performance applications. The center is staffed by more than 100 experienced technicians who oversee the techniques department, product development department, material analysis lab, and performance testing lab. These departments and labs work together to conduct research on new materials and techniques, test battery performance, inspect products, and test the performance of machines used in manufacturing.

3.2.9.2. Manufacturing

The manufacture of rechargeable batteries requires coordinated use of machinery and raw materials at various stages of manufacture. Shenzhen Highpower Technology has a large-scale production base that includes a 487,756-ft^2 factory; dedicated design, sales, and marketing team; and about 2,174 company-trained employees. It uses automated machinery to process key aspects of the manufacturing process to ensure high uniformity and precision, while leaving the non-key aspects of the manufacturing process to manual labor. It intends to improve its automated production lines and invest in its manufacturing infrastructure to further increase its manufacturing capacity and thus control the per-unit cost of its products.

3.2.9.3. Quality Control

Shenzhen Highpower Technology considers QC an important business practice. It has stringent QC systems that are implemented by more than 100 company-trained staff members to ensure QC in each phase of the production process, from the purchase of raw materials through manufacturing. Supported by advanced equipment, it uses a scientific management system and precision inspection measurements to produce stable, high-quality rechargeable batteries. Its QC department executes the following functions:

- Sets internal controls and regulations for semifinished and finished products,
- Tests samples of raw materials from suppliers,
- Implements sampling systems and sample files,
- Maintains the quality of the equipment and instruments, and
- Articulates the responsibilities of the QC staff.

The company monitors quality and reliability in accordance with the requirements of Quality System Review (QSR) and ISO 9001 systems. It has received EU CE attestation, UL authentication, and ISO 9001:2000 and ISO 14001 certification. It has passed stringent quality reviews and met original equipment manufacturer (OEM) qualifications of various domestic cell phone brands. Because of the company's technological capabilities and use of automated equipment for core aspects of the manufacturing process, the quality of its products meets and, in some key aspects, exceeds international industry standards.

3.2.9.4. Sales and Marketing

Shenzhen Highpower Technology has a broad sales network in China and one branch office in Hong Kong. The office sales staffs target key customers by arranging in-person sales presentations and providing post-sales services. They work closely with customers to address their needs and improve the quality and features of the company's products.

3.2.10. Shenzhen Wisewod Technology Co., Ltd.

Shenzhen Wisewod Technology Co., Ltd., is located at C Spot, Industrial City Area, Liantang, Gongming, BaoAn District, Shenzhen City, China. Its web site is www.wisewod.com. Ninety percent of the company's batteries are for cell phones. It is doing research on power batteries for electric bikes. The capacity for Li-ion battery packs for E-bikes is 7–12 A·h, and they are mainly 24 V. The positive materials are $LiMn2O4$ and $LiFePO_4$. The price of an E-bike lead acid battery pack is 300 ¥, and the whole bike costs 1,100–1,200 ¥; an Li-ion battery costs 1,300 ¥.

Shenzhen Wisewod Technology is a high-tech enterprise that produces and sells various Li ion batteries. Its workshops cover 40,000 m^2, and it can produce up to 600,000 Li ion batteries daily. Since it was established, the company has been focusing on product quality. It is determined to obtain a reputation for high quality and a performance-to-price ratio that will meet its customers' needs.

The company has a group of high-quality management and R&D personnel who pursue scientific innovation and excellence and have an enterprising spirit. Guided by an ISO 9001 quality management system, the company plans to open new business areas and take them to an international market.

3.2.11. Shenzhen Herewin Technology Co., Ltd.

Shenzhen Herewin Technology Co., Ltd., is located at Block A2, Haohaihong Garden, 4th Industrial Zone, Republic Village, Shajing Town, Bao'an District, Shenzhen. The company is making Li-ion battery packs in a LiFePO4 series and LiMn2O4 series for electric bikes. They are not yet being mass produced. Tables 3-4 and 3-5 provide specifications on the two series.

Shenzhen Herewin Technology is a high-tech company that was established in mid-2004 to develop, manufacture, and market Li-ion polymer battery cells and batteries in the south part of China. These batteries can be applied for sea transportation and fast information exchange. The company covers 30,000 m^2 and has a running capital of RMB 60,000,000.

Its Li-ion polymer battery products received UL, CE, and SGS (SGS S.A. or SGS Group, originally Société Genéralé de Surveillance) approval and ISO 9001:2000 international quality certification. The company is also honored as a "Shenzhen High-Tech Enterprise" by the Shenzhen Technology Bureau.

With the help of its co-founder, Central South University of Technology, Shenzhen Herewin Technology has built a strong, stable R&D team composed of 110 professionals, most of whom have a Bachelor's degree or higher. They are developing practical new products. As a result of the company's continuous research and marketing efforts, the range of its products has expanded into three high-temperature and high-capacity series. The products are widely used to power electronics, such as portable DVD players, PDAs, MP3 and MP4 players, radio control models, and power tools, as well as E-bikes and EVs. In the future, the company plans to work on more advanced techniques to make high-quality, competitively priced products.

Table 3-4. Specifications for Shenzhen Herewin Technology LiFePO$_4$ Li-Ion Polymer Batteries.

Parameter	24 V, 10 A·h	36 V, 10 A·h	48 V, 10 A·h	48 V, 20 A·h
Size (mm)				
Length	100	140	180	350
Width	100	100	100	100
Height	170	180	180	180
Capacity (A·h) at 0.2 C	10	10	10	20
Charging time (h)				
0.2 C (standard)	6–7	6–7	6–7	6–7
0.5 C (rapid)	2–3	2–3	2–3	2–3
Impedance[a] (mC)	≤200	≤250	≤300	≤300
Weight (kg)	~2.2	~3	~4.2	~8.5
Current (A)	5–20	5–30	5–40	5–50
Carriage (kg)	75	75	75	75
Cycle life (no. of cycles)	~1,000	~1,000	~1,000	~1,000

[a] Power control module (PCM) included.

Table 3-5. Specifications for Shenzhen Herewin Technology LiMn$_2$O$_4$ Li-Ion Polymer Batteries.

Parameter	24 V, 10 A·h	36 V, 10 A·h	48 V, 10 A·h	48 V, 20 A·h
Size (mm)				
Length	90	120	150	320
Width	100	100	100	100
Height	170	180	180	180
Capacity (A·h) at 0.2 C	10	10	10	20
Charging time (h)				
0.2 C (standard)	6–7	6-7	6–7	6–7
0.5 C (rapid)	2–3	2–3	2–3	2–3
Impedance[a] (m)	≤200	≤250	≤300	≤300
Weight (kg)	~1.8	~2.7	~3.5	~7.0
Current (A)	5-20	5-30	5-40	5-50
Carriage (kg)	75	75	75	75
Cycle life (no. of cycles)	~500	~500	~500	~500

[a] PCM included.

3.2.12. Shenzhen Bo Yi Neng Co., Ltd.

Shenzhen Bo Yi Neng Co., Ltd. (BYN) is located in Shen Zhen City, China. BYN was founded in 2006 with only 100 employees. Its power batteries are mainly for flying-vehicle-type toys. BYN conducts R&D on and manufactures polymer Li-ion batteries for a broad range of applications, such as mobile phones, Bluetooth earphones, MP3 and MP4 players, IPods, digital cameras, laptops, mobile DVD players, electric tools, and E-bikes. Its Li-ion batteries for E-bikes are not yet mass produced.

The company was founded by Dr. Zicai Zhou, a pioneer in Li-ion battery R&D in China. Under his leadership, the company pursued quality while advocating innovation. It has talented engineering and dedicated service teams. Since its inception, it has created and adopted new technologies in its design and manufacturing. Its products have received ROHS, TÜV Rheinland Group (TÜV), ISO 9001: 2000, and UL certification. They are sold in mainland China, Taiwan, Southeast Asia, Europe, and North America. Its quality and service have earned the trust of customers all over the world. Table 3-6 gives specifications on its batteries.

3.2.13. Aluminum Corporation of China

Aluminum Corporation of China (Chinalco) is an investment management and holding company owned and authorized by the state and under the direct administration of the central government. At the end of 2007, its total assets were more than 200 billion ¥ ($28.6 billion). It is the world's second-largest producer of alumina and third-largest producer of primary aluminum. With Chinalco being the holding company, Aluminum Corporation of China Ltd. (CHALCO) is listed on the Hong Kong, New York, and Shanghai stock exchanges. CHALCO was rated BBB+ by Standard & Poor's for 3 years.

Positioned to build itself into an international polymetallic mineral company, Chinalco is integrating its domestic resources and accelerating expansion of its global business with a wide- ranging product portfolio. The company is engaged in business based on laws in the following areas:

Table 3-6. Specifications for BYN Batteries.

Parameter	Lead Acid	Ni-Cd	Ni-MH	Li Ion	Li Polymer
Safety	Good	Good	Good	Good	Excellent
Nominal voltage (V)	2	1.2	1.2	3.7	3.7
Gravimetric energy density (Wh/kg)	35	41	50–80	120–160	140–180
Volumetric energy density (Wh/L)	80	120	100–200	200–280	>320
Cycle life (no. of cycles)	300	300	500	>500	>500
Operating temperature (°C)	−20 to ~60	20 to ~60	20 to ~60	0 to ~60	0 to ~4
Memory effect	No	Yes	No	No	No
Self-discharge (%/month)	<0	<10	<30	<5	<5
Environmentally friendly	No	No	No	No	Yes
Design form factor	None	None	None	None	Yes

- Investment and operation management of state-owned assets;
- Mining and beneficiation of bauxite, alumina refining and aluminum smelting, downstream fabrication, and trading;
- Mining and ore-dressing, smelting, processing, and trading of rare metals and rare earth;
- Mining, smelting, processing, and trading of copper and other nonferrous metals; and
- Related engineering and technological services.

Chinalco has independently developed and applied new technologies, such as the ore-dressing Bayer process and the 400-kA high-amperage electrolytic aluminum cell. The special aluminum alloys and titanium alloys produced by the company have become the key materials of the national defense industry and have been used in lunar launch vehicles and the Shenzhou spacecraft. The company has an outstanding management team and a technical expert team that deal with a full range of technological areas. At present, it owns many core technologies with independent intellectual property rights.

Embodying a corporate spirit that "strives for strength and excellence through hard work and innovation," a management style that is "strict, meticulous, pragmatic, innovative, persistent, and united," and a philosophy whose goal is to "maximize returns through honest operation," Chinalco promotes reform and development while honoring its social responsibility to be resource-efficient, environmentally friendly, and safe. Facing a new economic situation and global competition, Chinalco is determined to become a leading, stable, and profitable company, to realize sound and rapid growth through good science, excellent performance, and a well- developed corporate culture.

3.2.14. Tongji University

The School of Automobile Engineering was formally established in the Shanghai International Auto City in the Jiading District in 2002. It was a result of the merger of the Automotive Engineering Department, New Energy Center of Automotive Engineering, and the College of Automobile Marketing and Management, in accordance with the requirements of the Shanghai automotive industry. Now it is a college in Tongji University. It has a staff of 64, of which 19 are full professors, 16 are associate professors, and 13 are lecturers. It has 730 full-time undergraduate students, 124 master's degree students, and 38 doctoral students. There are 15 postdoctoral researchers in the mobile research center. In addition, the college has set up an internship program at the master's-degree level with several automobile companies. The college has extensive collaborative programs with several universities in Germany and the United States.

Dr. Wei, a professor at Tongji, provided a copy of the report, *Market Research on Power Li-Ion Battery in China 2007,* prepared by the China Social Economic Investigation & Research Center. A summary of this report is provided in Appendix F.

Drs. Sun and Wei at Tongji are interested in cooperating with U.S. companies and institutions. They would like to work with U.S. national laboratories on these topics as a starting point:

- Battery management technologies,
- Battery and vehicle modeling,
- Training Chinese students in the United States, and
- Battery hardware-in-the-loop testing and evaluation.

Tongji University has cooperative programs with institutions in Germany. Drs. Sun and Wei would like to apply that model to institutions in the United States. They are also interested in having a joint technical battery technology workshop and meeting in China every two years. Dr. Wei will submit a formal proposal on this in the near future.

3.2.15. DLG Battery (Shanghai) Company, Ltd.

DLG Battery (Shanghai) Company, Ltd. (DLG), was founded in 2001. It is situated in the most active and competitive economic area of China — Shanghai High-Tech Industrial Park. DLG occupies an area of 36,000 m^2 and employs 500 people, including 36 R&D engineers and 34 QC specialists. DLG makes 30,000 Li-ion cylindrical batteries, 20,000 Li-polymer prismatic batteries, and 3,000 Li-ion batteries per day. These batteries are provided to suppliers for computer notebooks, video camcorders, digital cameras, telephones, electric bikes, power tools, wheelchairs, and EVs. Its total capital is 150 million ¥ ($22 million). The production of Li cells is both semiautomatic and manual. DLG receives R&D funding from the (a) Shanghai Innovation Fund for Technology-Based Firms, (b) National Innovation Fund for Technology- Based Firms, (c) Shanghai Industrialization Fund for High-Tech Projects, and (d) Australian Research Council; the amount is not known. R&D projects include the following:

- High-rate-discharge polymer Li-ion batteries used in radio-controlled models and toys;
- High-rate-discharge cylindrical Li-ion batteries used in power tools;
- Power LiFePO$_4$ batteries used in E-bikes, EVs, and UPSs (in cooperation with K2 Company in the United States);
- High-capacity cylindrical Li-ion batteries used in digital equipment; and
- Miniature Li-ion batteries for implantable medical devices (in cooperation with the University of Wollongong, Australia).

DLG has several patents in China, including these:

- Cylindrical Li-ion battery with low polarization and a long cycle life (Patent No. ZL2004 2 0036999.7),
- Cylindrical Li-ion battery with safety valve controlled by temperature and pressure (Patent Application No. 2005 1 0023217.5),
- Cylindrical Li-ion battery for a digital camera (Patent No. ZL2005 2 0038860.0),
- Cylindrical Li-ion battery with internal PCM (Patent No. 2005 2 00468 15.x),
- A high-rate cylindrical Li-ion battery (Patent No. 2006 2 0041590.3),
- A novel method for manufacturing the cathode for a high-rate Li-ion battery (Patent No. 2006 1 0026342.6), and
- A sealing method and equipment for a power polymer Li-ion battery (Patent Application No. 2006 1 0027142.2).

Since its inception, DLG has focused on world-class quality and innovation. It has obtained ISO 9001-2000, CE, UL, and ISO 14001 certification. It has captured a share of both the international and domestic market. DLG is supplying batteries to such corporations as LG, GE, Hantel, Great Star, Toplink, Mosta Power Tools, EMAX, and K2 Solutions in the United States. DLG's business is 60% export and 40% domestic.

Figures 3-1 through 3-3 provide data and performance curves for DLG's cells and batteries. The batteries in Figure 3-1 are used in notebooks, video cameras, digital cameras, various portable equipment, and telephones. The batteries in Figure 3-2 are used in scooters, wheelchairs, and E-bikes. The batteries in Figure 3-3 are used in power tools and EVs.

DLG has been paying close attention to QC and customer needs and is working on providing these after-sales services to customers:

- Solve customer complaints (respond to complaint within 24 hours, provide a report to the customer within 1 week),
- Answer customer questions,

High Rate Discharge Li-Polymer Battery | 高倍率聚合物锂离子电池系列

Model 型号	Rated Capacity 额定容量(mAh)	Thickness 厚度(mm)±0.3	Width 宽度(mm)±0.5	Height 高度(mm)±0.5	Weight 重量(g)	Impedance 内阻(mΩ)	Nominal Voltage 额定电压(V)	Discharge rate 放电倍率
PLB401218H-003	30	4.2	12	19	1.1	300	3.7	6C
PLB501417H-005	50	5.8	14	18	1.5	240	3.7	10C
PLB701417H-007	70	7.0	14	18	2.1	240	3.7	10C
PLB401233H-007	70	4.2	12	34	2.2	140	3.7	10C
PLB452026H-013	130	4.6	20	27.5	3.5	110	3.7	10C
PLB402030H-015	150	4.0	20	30.5	5.5	120	3.7	10C
PLB452030H-018	180	4.5	20	30.5	6.0	100	3.7	10C
PLB602030H-025	250	6.0	20	30.5	7.0	80	3.7	10C
PLB403048H-035	350	4.0	30	48.5	11.5	60	3.7	10C
PLB603048H-050	500	6.0	30	48.5	16	60	3.7	20C
PLB703048H-080	800	7.0	30	48	20	40	3.7	12C
PLB853048H-085	850	8.6	30	48	22	15	3.7	15C
PLB503450H-060	600	5.0	34	50	17	50	3.7	12C
PLB653562H-105	1050	6.5	35	62	28	40	3.7	12C
PLB803562H-130	1300	8.0	35	62	33	30	3.7	12C
PLB803562H-125	1250	8.0	35	62	32	10	3.7	15C
PLB803480H-160	1600	8.0	34	80	42	25	3.7	12C
PLB603496H-160	1500	6.0	34	96	38	40	3.7	12C
PLB653496H-155	1550	6.5	34	96	40	10	3.7	15C
PLB553496H-130	1300	5.5	34	96	38	12	3.7	15C
PLB753496H-180	1800	7.5	34	96	46	10	3.7	15C
PLB803496H-200	2000	8.0	34	96	50	8	3.7	15C
PLB853496H-220	2200	8.5	34	96	52	8	3.7	15C
PLB703496H-165	1700	7.0	34	96	46	8	3.7	15C
PLB703496H-150	1500	7.0	34	96	45	10	3.7	25C
PLB853496H-200	2000	8.4	34	96	53	8	3.7	25C
PLB803496H-180	1800	8.0	34	96	50	8	3.7	25C
PLB903496H-220	2200	8.6	34	96	55	6	3.7	25C
PLB8548135H-500	5000	8.5	48	135	120	15	3.7	10C

Performance Curve | 性能曲线

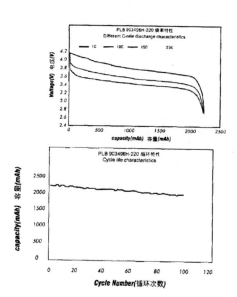

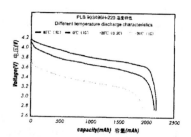

High rate discharge performance
Excellent safety performance
Multiple options
Light weight
Low internal resistance
Full range products for RC toys/helicopters/aircrafts

Figure 3-1. Performance of DLG High-Rate-Discharge Li-Polymer Battery.

High Energy Cylindrical LiFePO₄ Battery | 高能量型磷酸铁锂电池系列

Model 型号	Rated Capacity 额定容量(mAh)	Diameter 直径(mm)±0.2	Height 高度(mm)±0.5	Weight 重量(g)	Impedance 内阻(mΩ)	Nominal Voltage 额定电压(V)	Discharge Rate 放电倍率
IFR16340-050(J)	500	16.5	33.7	17	≤150	3.2	4C(with PTC)
IFR18650-140	1400	18.2	64.5	40	≤80	3.2	4C
IFR26650-280	2800	26	65	76	≤40	3.2	4C
IFR26650-320	3200	26	65	78	≤40	3.2	4C
PFB 9V	220	48.5(±1) X 27(±1) X 15.5(±1)		30	≤1000	9.6	1C

Performance Curve | 性能曲线

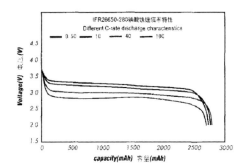
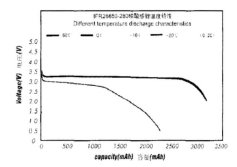

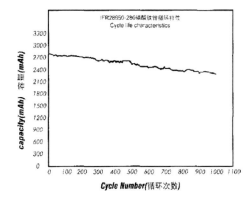

Better safety, better stability
Rapid charging performance
Long cycle life
For EB/EV/UPS
Environment friendly

Figure 3-2. Performance of DLG High-Energy Cylindrical LiFePO4 Battery.

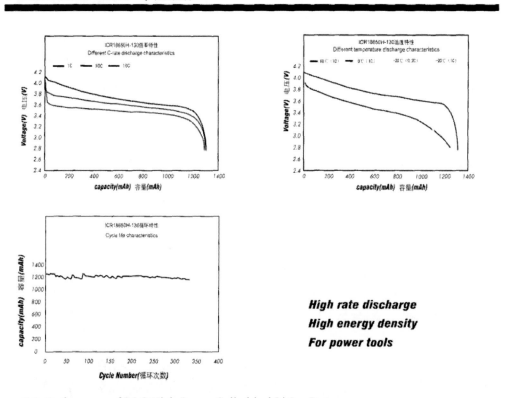

Figure 3-3. Performance of DLG High-Power Cylindrical Li-Ion Battery.

- Provide technical reports to customers after the DLG technical department has finished testing,
- Offer technical and equipment-related material to customers,
- Provide training on battery characteristics and uses,
- Investigate customer satisfaction each quarter, and
- Evaluate each customer's request and ensure that the request is addressed.

This list indicates that Chinese companies that are developing advanced battery technologies are also beginning to pay attention to quality and customer satisfaction. This is a big step forward for them, and as a result of this focus, they hope to get a larger share of the international market and play a leading role in developing battery technologies.

3.2.16. A-SI-KA Electric Bike Co., Ltd.

A-SI-KA Electric Bike Co., Ltd. (ASK), purchases Li-ion batteries from several battery companies. It buys a major share from DLG. ASK has three sizes of batteries: (1) 36 V and 12 A•h with a 30-km range, (2) 48 V and 10 A•h with a 40-km range, and (3) 24 V and 3.2 A•h with a 20-km range. Batteries cost 2,600–2,000 ¥ (U.S. $370–285). In comparison, lead acid batteries cost $1,050 ¥ (U.S. $150). Also, lead acid batteries last for about 2 years, while ASK expects Li-ion batteries to last for 4–5 years.

ASK makes several different sizes of bikes, including a folding bike. One feature of the bikes is that when a bike is parked, the owner can remove and keep the battery. This feature not only prevents thefts, it prevents riders from becoming stranded without a battery. Figure 3-4 provides a photo of a bike; the technical parameters and specifications for a 16-in., 36-V bike; and a strategy for troubleshooting problems.

ASK prepared a step-by-step user manual in Chinese and English that provides (1) information on bike safety requirements, operation, and precautions and (2) instructions on charging the battery and caring for and maintaining the bikes. In case there is a malfunction, the manual also describes various problems, their causes, and troubleshooting methods.

3.2.17. Wanxiang Group

Wanxiang Group was established in 1969. At first, it had only seven employees and 4,000 ¥ (U.S. $570) in assets. After more than 30 years of development and innovation, it is now listed as one of 120 pilot enterprise groups by the Hangzhou state council and one of 520 national key enterprises, and it has more than 20,000 employees and 10 billion ¥ in assets. Wanxiang Group is one of China's leading manufacturers of auto parts. With a goal of integrating into the global economy, Wanxiang Group manufactures automobile parts for the main world markets and advanced technologies for an international arena. It is expected to strengthen the development of the high-tech industry, gradually becoming a multinational group that complies with international routines.

The overall goal of Wanxiang Group is to become a modern company while retaining core values. Its latest near-term goal was to make a profit of 10,000,000 ¥ (U.S. $1,428,571) by the end of 2008 and provide its employees with an annual income of 6,000,000 ¥.

3.2.17.1. Wanxiang Electrical Vehicle Co., Ltd.

Wanxiang Electrical Vehicle Co., Ltd. (WEVC), founded in 2002, is a subcompany controlled entirely by Wanxiang Group. WEVC focuses on developing clean energy technologies and energy-saving technologies, such as electric passenger vehicles (Figure 3-5) and buses. WEVC's achievements in developing high-power and polymer Li batteries, integrated motor and drive-control systems, whole-vehicle electronic control systems, vehicle engineering integration techniques, and test trial platforms have been outstanding. In the challenge known as "Bibendum," WEVC won the overall contest prize for appearance and EV performance given by the International Vehicle Association.

Main Technical Parameters and Specification (16" 36V)

Entire Bicycle		Motor	
Weight:	19kg	Type: High-efficiency DC Permanent-magnet Brushless Motor	
Load Capacity	75kg	Maximum driving noise:	≤62db
Size:	1410×540×1040mm	Rated power:	180W
Maximum mph:	20km	Rated speed:	235r/min
Driving distance after charge:	≥30km	Rated Voltage:	36V
Climbing Ability:	≤12°	Rated efficiency:	≥78%
Over current protection value:	15±1A	Power consumption per 100 kilometer	≤1.2kw.h/100km
Under voltage Protection value:	30V	Weight: (KG)	≤2kg

Battery		Charger	
Type:	lithium	Input Voltage:	AC110~220V50Hz
Voltage:	36V	Rated Output:	42V/1.5A
Capacity:	8/9 Ah	Charging Time:	5－8h

Illustrative Drawing for the Vehicle Electricity & Common troubles and Troubleshooting

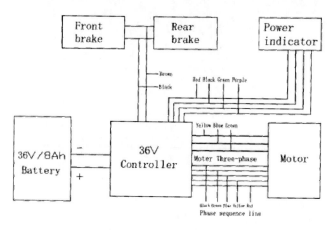

Figure 3-4. Photograph, Technical Specifications, and Troubleshooting Details for an ASK Bike.

Since 1999, Wanxiang Group and Zhejiang Province have been engaging in R&D on EVs. They established Wanxiang Electrical Company to develop a "new energy vehicle," and the company has been accumulating the necessary technology to do so. So far, WEVC is the only company with the key technology for a battery/motor/electrical-control/power-plant system. WEVC was charged with developing five items related to the National 863 Plan project and four items important to the Zhejiang Province science and technology project. Wanxiang Group owns more than 50 patents. In its Y9 "pure" electric bus (Figure 3-6), WEVC installed a polymer Li battery and power system that ran more than 300,000 km around Hangzhou West Lake, demonstrating its reliability. In 2006, the National 863 Plan new energy vehicle project, "Research and Development of a Pure Electrical Power System Platform for a Passenger Vehicle," was assigned to WEVC. This assignment was evidence of the company's leading position in the EV industry. It has become a pioneer in the state's pure electric vehicle development project. WEVC has several types of polymer Li batteries, from a 24-V, 8-A•h battery for E-bikes to a 600-V, 400-A•h battery for a pure electric bus. It also has a power system platform that could supply a pure electric car, pure electric bus, dual-energy trolley bus, or HEV bus (Figure 3-7) with power from 3 to 150 kW.

Technical Feature:

(1) Adopting Pure Electrical Vehicle Platform;
(2) Module Technology "Common Platform shared by Several Models"
(3) Supplied with Polymer Lithium Hydronium Power Battery Piles;
(4) Charging on-Board or Adopting Structure with fast-replace Battery piles;
(5) AC Driving System with Advanced Vector Control Technology;
(6) Adopting Anti-Electric dizzy 3-phase AC Asynchronism Motor with Aluminium Structure;
(7) Frequency-Conversion Controller with Advanced Vector Controlling Technology;
(8) AC supplement system (Air-condition, Steering and Braking) With high efficiency and energy-saving;

Technical Data:

Specification	Parameters
Length×Width×High	4295×1705×1570 mm
	1520 Kg
Max. Gross Mass	1950 Kg
Battery Pile Power Battery Type	Polymer Lithium Hydronium Battery
Specification	88×100Ah in series
Driving Speed (Continuous/Peak)	4000/12000r/min
System Torque (Continuous/Peak)	210/900N•m
Power (Continuous/Peak)	17/37 Kw
Max. efficiency of Controller	≥98%
Range (50Km/h)	380 Km
Electric consumption ratio	9.8Wh/100 Km
Max. Speed	126Km/h
Accelerating time for speed 0–50km/h	9.2 s
Max. gradient	≥18%

Figure 3-5. Technical Features and Data for Electric Vehicles.

Technical Feature:

(1) Supplied with Polymer Lithium Hydronium Power Battery Piles;
(2) Adopting Dual-Motor Driving technology;
(3) Digital CAN network communication(including electric, power system and digital instrument);
(4) AC transmission control system with direct torque/space vector controlling technology;
(5) Adopting Anti-Electric dizzy 3-phase water-cooling AC Asynchronism Motor
(6) Battery management system in-Module;
(7) Frequency-Conversion Controller with Advanced Vector Controlling Technology;
(8) AC supplement system (Air-condition, Steering and Braking)With high efficiency and energy-saving;
(9) Adopting Structure with fast-replace Battery piles;

Technical Data:

Specification		Parameters
Length×Width×High		10860×2490×3460 mm
Max. Gross Mass		18600 Kg
Battery Pile	Power Battery Type	Polymer Lithium Hydronium Battery
	Specification	88×100Ah in series, 6 Parallels
Driving System	Speed (Continuous/Peak)	2000 rpm
	Power (Continuous/Peak)	2×45 Kw
	Max. efficiency of Controller	96%
Range		280Km
Max. Speed		90Km/h
Accelerating time for speed 0—50km/h		42s
Max. gradient		≥20%

Figure 3-6. Technical Features and Data for Electric Buses.

Technical Feature:

(1) Several energy controlling technology for public-transportation running-cases with Hybrid Power Coupling technology;
(2) "Engine+ ISG+ Electric control Clutch+ Driving Motor" integrated design;
(3) Supplied with Polymer Lithium Hydronium Power Battery Piles;
(4) Digital CAN network communication(including electric, power system and digital instrument);
(5) AC transmission controlling system adopting direct torque/space vector control technology;
(6) Whole vehicle control system based on DSP to control energy distribution and energy coupling;
(7) Adopting Anti-Electric dizzy 3-phase water-cooling AC Asynchronism Motor;
(8) Frequency-Conversion Controller with Advanced Vector Controlling Technology;
(9) AC supplement system (Air-condition, Steering and Braking)With high efficiency and energy-saving;

Technical Data:

Specification		Parameters
Length×Width×High		11850×2490×3340 mm
Kerb Weight		12980 kg
Max. Gross Mass		17980 kg
Diesel Engine		ISDE180 4 cylinders
ISG motor system	Rating RPM	3000 rpm
	Max. generating Power	50 kw
Battery Pile	Power Battery Type	Polymer Lithium Hydronium Bat
	Specification	84×100Ah in series
Driving System	Speed (Continuous/Peak)	1800rpm
	Power (Continuous/Peak)	63/120Kw
	Torque (Continuous/Peak)	802/1900N·m
	Max. efficiency of Controller	98%
Max. Speed		90 km/h
Accelerating time for speed 0—50km/h		≤12 s
Max. gradient		≥20%
Fuel consumption ratio vs reduction percentage		25%
Driving Range in pure electric mode		50 km

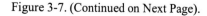

Figure 3-7. (Continued on Next Page).

Figure 3-7. Technical Features and Data for Hybrid Electric Buses.

3.2.17.2. Polymer Li-Ion Battery

The polymer Li-ion battery (Figure 3-8) is commonly called a gel Li-ion battery or plastic battery. Because of its flexible shape, it could also be called a gum battery. It can be coiled, cut, or folded easily. It is popularly recognized because of its high specific energy, excellent safety, flexible shape, long life, and the advanced and environmentally friendly process used to fabricate it. Polymer Li-ion batteries, which are much safer than liquid Li-ion batteries because there is no liquid electrolyte inside them, are the most promising batteries in the world. WEVC has more than 5 years of experience in developing plastic batteries ranging from 8 to 100 A·h. Its batteries employ the most advanced technology, have the best performance, and can be used in a wide variety of applications. Because of their safety, the batteries can be produced at both a technical- research scale and mass-production scale on electrolyte and cathode materials. The products are widely used in E-bicycles, automobiles, and UPS batteries. Their features are as follows:

- The energy-type lithium polymer cell has a maximum capacity of 100 A·h, energy density of 125 Wh/kg, and lifespan of 1,000 cycles at 80% depth of discharge (DOD).
- The power-type lithium polymer cell has a power density of up to 1,300 W/kg, energy density of 80 Wh/kg, and lifespan of more than 1,000 cycles.
- The products are certified by national authority. The batteries will not start on fire or explode under extreme conditions (e.g., if they are crushed or penetrated or if there is a short circuit).
- The total solution for a 24–310-V power source system can be provided.

3.2.17.3. Motor and Drive System

WEVC develops technologies to make standardized products that can be used in various EVs. Standardized products keep vehicle costs down. After 6 years of conducting technology research and collecting test data, WEVC acquired the key EV technology and assumed the leading position in the EV market. WEVC provided cooperating OEMs with the electric drive systems and energy systems they needed to develop various EVs. WEVC also develops different types of EV energy systems to reduce the degree, amount, and cost of energy used. WEVC uses these technologies in new EV and bus demonstration projects. In 2006, a 40-kW pure electric mini-bus was exported overseas. WEVC's motor and drive-system product covers a range of 3– 150 kW and can be used in electric cars and buses. Development of the motor and drive system is moving toward forming various types of energy platforms and offering integrated solutions to improve vehicle acceleration and economics.

Characteristics:

1 High Energy Density Power Battery: Realizing mass production and sales Cell Capacity at the range from 8Ah to 100Ah, gravimetric energy density 125Wh/kg, volumetric energy density 260Wh/L, cycle life 1000 times (80%DOD).

2 High Rate Power Battery: power density 1300W/kg, gravimetric energy density 125Wh/kg, cycle life 1000 times (80%DOD).

3 Safety Performance: Power battery can't catch fire or explode in extreme circumstances, such as penetration, crush, overcharge or short-circuit etc.. The products are certified by National Authority, CE Certification, UL Certification.

4 Power Source Solution: various specification. 24V~310V/8~600Ah electrical bicycle, electrical motor electrical vehicle power sources have been realized industrialization.

Technical Data:

Specification	24V/8Ah	36V/10Ah	48V/20Ah	48V/30Ah	48V/50Ah	48V/100Ah	310V/100Ah	310V/600Ah
Rated Capacity/Ah	8	12	25	30	50	100	100	600
Minimum Capacity/Ah	7.5	11.5	24	28	48	95	95	570
Rated Voltage/V	25.9	37	48.1	48.1	48.1	48.1	310.8	310.8
Maximum Charge Voltage/V	29.4	42.0	54.6	54.6	54.6	54.6	352.8	352.8
Minimum Discharge Voltage/V	20.3	29	37.7	37.7	37.7	37.7	243.6	243.6
Charge Current/A	≤0.5C	≤0.5C	≤0.5C	≤0.5C	≤0.5C	≤0.5C	≤0.5C	≤0.5C
Rated Discharge Current/A	0.5C	0.5C	0.5C	0.5C	0.5C	0.5C	0.5C	0.5C
Maximum Discharge Current/A	2C	2C	2C	2C	2C	2C	2C	2C
Weight/kg	1.8	3.5	8.0	12.5	20	48	305	1820
Resistance/mΩ	≤60	≤70	≤70	≤50	≤50	≤40	≤110	≤50
Dimension Length/mm	245	310	260	260	265	265	1015*2	2030*4
Width/mm	84	95	150	150	160	160	870*2	870*4
Thickness/mm	60	82	150	200	480	910	245*2	245*4

Figure 3-8. Characteristics and Technical Data for Polymer Li-Ion Batteries.

3.2.17.4. Electric Control System

The electric control system is a pivotal technology for EVs. It ties all the independent subsystems into a complete system. It harmonizes the actions of the different subsystems. WEVC's Electric Control System Group mainly conducts research on constructing the digital communication network for EV bodies and on developing a new-style composite LCD meter. It also does research on control strategies, applied technologies for the battery management system, and the electric wiring safety system. To develop an electric control system, the direction being taken is to fabricate a digital network calculation platform for new-style automobiles. This platform should be able to conveniently supply excellent information for creating independent, economical, and environmentally friendly vehicles.

3.2.17.5. Vehicle Engineering

Vehicle engineering technology is the basis for integrating vehicle parts and components and developing next-generation vehicles that save energy and protect the environment. WEVC is making full use of the Wanxiang Group's advantages in vehicle parts development in its study of vehicle engineering technology, whose core components are chassis platform integration and vehicle body design. WEVC has been developing vehicle engineering technology in the following areas: three-dimensional digital clay modeling, computer-aided engineering (CAE) analysis of structural stress, vehicle dynamics simulation, vehicle performance match analysis, and prototype manufacturing and testing.

3.2.17.6. Market Development and Expansion

With a target of product industrialization, WEVC has been developing key EV components, such as a polymer Li battery, an integrated motor and control system, and a whole- vehicle digital-control system. WEVC has many development and industrialization capabilities: power battery development, CAE analysis of vehicle chassis design and structure, concept design and styling and body structure design, concept prototype development, EV proof of concept, and power system design.

With regard to development, demonstration, and industrialization of EVs, WEVC took a big step forward. Five of its electric buses ran more than 300,000 km around Hangzhou West Lake along Y9 routes, transporting more than 200,000 passengers. Expansion to 10 buses is expected. WEVC delivered two dual-energy trolley buses to Hangzhou Public Transportation Company. It also finished developing a parallel-series HEV (PSHEV) system (including a PHEV) and manufactured a prototype, exported a pure electric mini-bus to Taiwan, and sold polymer Li batteries with specifications from 24 V and 8 A•h to 600 V and 400 A•h.

3.3. Characteristics of Lithium-Ion Cells and Modules

Characteristics of Li-ion cells and modules for ETWs from various manufacturers are shown in Table 3-7.

3.4. Individual Battery Attributes

In China, the market for small-capacity Li-ion batteries for cell phones, laptops, and other devices is getting saturated. Large-capacity Li-ion power batteries have not entered the market yet. Li-ion power batteries are good enough to be used for electric bikes and electric motorcycles. The electric motorcycle made by Taiwan's EVT Electric Motorcycles Company employs a 3-V, 100-A•h Li-ion battery. It has a driving range of up to 200 km and speed of up to 100 km/h.

With regard to electric automobiles, solid polymer Li-ion batteries may play a main role in the coming 2–3 years. Their energy density is 30% higher than that of liquid Li-ion batteries, and their shape is more flexible; shapes could be strip-like, cylindrical, or prismatic.

With regard to electric motorcycles, there are three main types of power batteries: lead acid, Ni-MH, and Li-ion. So far, Ni-MH has been an ideal power source. Li-ion batteries have to overcome a few issues (e.g., take a fast charge and large current, balance cell voltages, and become safer) before they can be used in electric motorcycles. Requirements for protective electronic boards, a battery management system, and a thermal management system would increase the cost of Li-ion power batteries.

When EV production began in 1998, there were 16 companies, and annual production in China was 58,000. Since then, the EV industry has been developing very rapidly; annual growth has averaged 87%. In 2006, 19.5 million E-bikes were sold. The competitive pressure on E-bike companies has been increasing daily. When the number of bikes sold by a company in a year is fewer than 20,000–30,000, it is hard for it to survive.

The Li-ion power battery pack for an E-bike is composed of multiple unit cells in series or mixed in series and parallel. It is hard to maintain the consistency of each unit cell, and the inconsistency affects the safety and life of the whole battery pack. Thus, it is critical to keep the Li-ion batteries 100% safe if a good system for managing Li-ion battery packs is to be invented.

Table 3-7. Characteristics of Li-Ion Cells and Modules for ETWs from Various Manufacturers.

Mfgr.	Voltage and Capacity	Cathode Material	Nominal Impedance (mΩ)	Weight (kg)/ No. of Cells	Height	Weight	Length	Specific Energy (Wh/kg)	Cycle Life
MGL	24 V, 13 A·h	LiMn$_2$O$_4$	≤100	2.30	92	72	245	NA[a]	NA
	24 V, 20 A·h		≤100	3.940	95	105	280	NA	NA
	24 V, 8 A·h		≤100	2.450	60	90	260	NA	NA
Zhenglong Battery	12 V, 10 A·h	NA	NA	NA	95	99	151	200	>500
	37 V, 13 A·h		≤150	NA	NA	NA	NA	NA	NA
	48 V, 10 A·h		NA	NA	NA	NA	NA	NA	NA
Shenzhen Herewin	24 V, 10 A·h	LiFePO$_4$	≤200	2.2	180	100	100	NA	≥1,000
	36 V, 10 A·h		≤250	3	180	100	140	NA	
	48 V, 20 A·h		≤300	8.5	180	100	350	NA	
	24 V, 10 A·h	LiMn$_2$O$_4$	≤200	1.8	170	100	90	NA	≥ 500
	36 V, 10 A·h		≤250	2.7	180	100	120	NA	
	48 V, 20 A·h		≤300	7.0	180	100	320	NA	
Phylion Battery	24 V, 10 A·h	LiMn$_2$O$_4$	≤130	2.7/7 in series	135	67	171	NA	NA
	30 V, 10 A·h		≤140	3.3/8 in series	135	67	190	NA	NA
	37 V, 10 A·h		≤150	3.8/10 in series	135	67	228	NA	NA
Thunder Sky Li Battery	12 V, 40 A·h	LiFFePO$_4$	NA	1.6	46	116	188	NA	>2,000
	12 V, 40 A·h	LiFNiMnCoO$_4$	NA	1.6	46	116	188	NA	>300
China Powerel Battery	24 V, 10 A·h	NA	NA	2.2	95	95	325	NA	NA
	36 V, 10 A·h		NA	2.9	95	95	325	NA	NA
Hunan Yeshine King Co.	24 V, 10 A·h	LiFePO$_4$	≤100	3.0	88	57	270	NA	>2,000

[a] NA means not available.

$LiCoO_2$ is the main positive material for Li-ion batteries. Since 1990, Li-ion batteries have been commercialized in many developed countries, such as Japan, the United States, France, and Germany. In China, Li-ion batteries had been commercialized by the end of 20th century. Li-ion batteries with $LiCoO_2$ are being developed to have a longer life, higher capacity, and greater safety. The China $LiCoO_2$ battery industry has had to make a significant R&D effort for new products in order to be competitive and maintain sustainable development.

$LiMn_2O_4$ Li-ion batteries are being used in portable electronics, in communication and military equipment, and in transportation. They could be used as energy storage devices to complement the exploration of wind and solar energy.

$LiFePO_4$ Li-ion batteries are safe, perform well in high-temperature environments, have good capacity, and are low in cost (one-fourth the price of $LiCoO_2$ batteries). $LiFePO_4$ Li-ion batteries can be used in energy storage devices for solar and wind generator systems, UPSs, power tools, EVs, medical equipment, toys (remote-controlled electric toy planes, boats, vehicles, etc.), and other items.

With regard to Li-ion power batteries, $LiMn_2O_4$, $LiFePO_4$, and $LiMn_xNi_yCo_{1-x-y}O_2$ will share the market as positive materials in the coming 3 years. After 3 years, $LiFePO_4$ will occupy a much bigger market. In 3 years, it is estimated that the demand for $LiFePO_4$ batteries will reach more than 10,000 tons/yr.

So far, the main competitors to $LiFePO_4$ batteries in China have come from Valence Technology, Inc. (U.S.), A123 (U.S.), and Tianjin STL Energy. However, most $LiFePO_4$ companies (e.g., Huannan Reshine, Pulead Technology Industry Co., Ltd.) still have issues with regard to the stability of (consistency between) production batches.

3.5. Cost of Lithium-Ion Batteries for Electric Two-Wheelers

A Li-ion battery pack costs three to four times more than lead acid batteries. It was thought that an increase in the price of a ton of lead from 15,000 ¥ in 2006 to 25,000–34,000 ¥ in 2007 would improve the market for Li-ion batteries, but this large price increase did not happen. The price of an E-bike lead acid battery pack was about 300 ¥ in 2007, and the price once reached 720 ¥ for a 48-V, 12-A·h battery. In 2008, a bike lead acid battery cost 1,500–2,000 ¥, while a Li-ion battery cost about 2,300 ¥.

Table 3-8. Production of Secondary Batteries in China, 2002–2006.

Year	Lead Acid Battery Production (10^6 units)	Ni-MH Battery Production (10^6 unit cells)	Ni-MH Battery Exports (10^6 unit cells)	Ni-Cd Battery Exports (10^6 unit cells)
2002	500	Not available	Not available	Not available
2003	600	Not available	Not available	Not available
2004	750	Not available	Not available	Not available
2005	900	960	870	350
2006	1,005	1,100	960	430

3.6. Recent Battery Production in China

Table 3-8 shows data on the production of secondary batteries in 2002 through 2006 in China. The growth in demand for Li-ion power batteries in 2001 through 2007 in the global market was as follows: 8.3% in 2001, 9.2% in 2002, 10.1% in 2003, 11.6% in 2004, 12.7% in 2005, 13.6% in 2006, and 15.3% in 2007.

4. ELECTRIC TWO-WHEELERS IN CHINA

4.1. Introduction

China is still a developing country. Most Chinese people still have a low quality of life relative to that in other countries. For them, ETWs are good choices for transportation in terms of economy, convenience, and effectiveness. The vehicles also occupy less space on the road and are better for the environment.

In 2006, 20 million E-bikes were made in China. Currently, China has 50 million battery-operated bicycles on the road. Only a very small percentage of them operate on Li-ion batteries; the rest use lead acid batteries. About 2,500 companies in China produce electric two-wheeled or three-wheeled vehicles.

All the large EV companies have E-bike models that use Li-ion batteries, but their performance-to-price ratio is still not compatible with that of E-bikes with lead acid batteries. This is the key reason that bikes that use Li-ion batteries are still not in mass production. Despite this fact, it is popularly thought in China that Li-ion power batteries could rapidly replace lead acid and Ni-MIT batteries as main power sources for ETWs.

4.2. Chinese Electric Two-Wheeler Industry

4.2.1. Development of Electric Two-Wheelers in China

About 2,500 companies in China produce electric two-wheeled or three-wheeled vehicles. The main companies are distributed throughout Jiangsu Province, Zhejiang Province, and the city of Tianjin. They include Jiangsu Yadea Technical Development Co., Ltd.; Tianjin Fushida Electric Bicycle Co., Ltd.; Jiangsu Xinri Electric Bicycle Co., Ltd.; and Zhejiang Luyuan Electric Vehicle Co., Ltd. These companies produce more than 400,000 ETWs per year. The number of E-bike companies in Shandong Province and the city of Shanghai also increased rapidly during 2005 and 2006. The main Chinese company that has been developing high-end ETWs is Wuxi in Jiangsu Province, while Tianjin is still the top production base for simple ETWs.

It was predicted that the total amount of ETWs in China would be close to 20 million in 2008 and more than 30 million (including 5–6 million exports) in 2010. Total revenue could reach 70 billion ¥ in 2010.

Table 4-1 shows the number of number of automobiles, motorcycles, and E-bikes produced and on the road in China in 2001–2007. Table 4.2 shows the growth in demand for E-bikes and E-bike power batteries, in the production of E-bike batteries, and in exports of power tools in China in 2001–2007. In 2007, more than 20 companies that produced ETWs started to produce E-bikes with Li-ion battery packs, including 24-, 36-, and 48-V batteries.

4.2.2. Some Manufacturers of Electric Two-Wheelers in China

4.2.2.1. Jiangsu Yadea Technical Development Co., Ltd.

This company is a leader in China's electric bike and special vehicle industry. It has two branches, in Tianjin and Guangzhou. The headquarters in Wuxi, Jiangsu Province, is a large-scale base of production that occupies 500 acres. The company pursues innovation and progress, incorporates advanced technologies from around the world, and develops its own advanced technologies. The company is consumer oriented and has 567 service centers. In April 2007, it cooperated with Wan Lixing to represent China's EV industry at the international trade show in Beijing. It produces quality ETWs. Despite its focus on high-speed technological development, the company has not forgotten its debt to society. When a consumer purchases one of its vehicles, it contributes 2 ¥ to the Hope Project, a fund for children's education, and it plans to increase its support to help the project establish primary schools. Its target is to advance its brand name domestically within 3 years and internationally within 5 years.

4.2.2.2. Luyuan Electric Vehicle Co., Ltd.

This company, which was established in 1996, is located in Jinhua City, Zhejiang Province, China. It covers 130 acres and employs 1,600 workers. With 12 production lines, its annual production capacity is up to 1 million EVs. The company accounts for 5% of the Chinese EV market and distributes its products to more than 20 countries. Its output reached 200,000 vehicles in 2005, 300,000 in 2006, and 500,000 in 2007, and it was expected to be 600,000 in 2008.

Table 4-1. Number of Automobiles, Motorcycles, and E-Bikes Produced and on the Road in China, 2001–2007.

Year	No. of Cars Produced (10^6)	No. of Cars on the Road (10^6)	No. of Motorcycles Produced (10^6)	No. of Motorcycles on the Road (10^6)	No. of E-Bikes Produced (10^6)	No. of E-Bikes on the Road
2001	2.344	18.02	9.96	47.6	0.4	
2002	3.251	20.53	11.5	51.0	1.6	
2003	4.444	23.83	15.0	60.0	4.0	
2004	5.074	27.42	14.75	67.5	6.75	13,000
2005	5.708	35.0	17.24	76.3	12.11	23,000
2006	7.280	41.0	21.45	83.5	19.5	37,500
2007 (prediction)	8.50	47.5	25.4	94.0		

Based on the report by the China Social Economic Investigation & Research Center, *Market Research on Power Li-Ion Battery in 2007, China*.

Table 4-2. Growth in Demand for E-Bikes and E-Bike Batteries, Production of E-Bike Batteries, and Exports of Power Tools in China, 2001–2007.

Year	Growth in Demand for E-Bikes (%)	Growth in Demand for E-Bike Batteries (%)	Growth in Production of E-Bike Batteries (%)	Growth in Exports of Power Tools (%)
2001	36.5	41	32.3	10.2
2002	300	282	216.4	11.7
2003	150	163	137.5	12.8
2004	68.8	65	58	13.6
2005	79.4	81	76	15.9
2006	61	59	67	17.6
2007 (prediction)	59			
2008 (prediction)	63			

Based on the report by China Social Economic Investigation & Research Center, *Market Research on Power Li-Ion Battery in 2007, China*.

4.2.2.3. Shandong Incalcu Group Co., Ltd.

This company is a large-scale group enterprise that integrates scientific research, manufacturing, and trade. With its modern scientific management system and the technological progress it has achieved, it is one of 200 key enterprises in Shandong Province. It has six whole-asset subsidiaries and four shareholding companies. The Incalcu brand name has been a "Shandong Province Famous Brand" since 1997, and it won awards from the Chinese National Government as a "China Famous Brand" in November 2004 and as "China's Top Brand" in September 2006. The company's products passed ISO 9001 QC certification. Its main product is exercise equipment, and it also makes E-bikes, electrode aluminum foil, and light-electric products, among others. The products are sold in the domestic market and Japan, the United States, EU, Middle East, Southeast Asia, and South America.

4.2.2.4. Yangzhou Feichi Group Corp.

This corporation, located in the ancient city of Yangzhou, is a strong manufacturer of EVs and ETWs. Its has about 15,900 m^2 of workshops that occupy a beautiful environment of 50,000 m^2 and employs more than 60 technicians who are university graduates. It is a modern enterprise where scientific research, production, and trade are synthesized. The corporation has several affiliated companies, including electric bicycle, electric appliance, and dynamic mechanism companies. Feichi Electric Bicycle Co., Ltd., produces electric bicycles, as well as electric sliding boards and electric tricycles. Its products have been sold to many countries in Europe and in America, and it has established agents in many Chinese cities. The Feichi brand has become a favorite to customers throughout China. Feichi Electric Appliance Co., Ltd., produces home-use kitchen garbage processors, which have recently been exported on a large scale. Feichi Dynamic Mechanism Co., Ltd., produces hydraulic brake discs for motorcycles; they are known all over the world, having become the base for Japanese Suzuki,

Honda, and Yamaha vehicles. Feichi products are now being promoted with great enthusiasm throughout the world in order to leverage the entrance of China into the World Trade Organization. "Being people who can contribute to society" is an aim of all Feichi employees, and a company target is to contribute to environmental protection.

4.2.2.5. Wuxi Kawamura Bicycle Co., Ltd.

This Sino-Japanese joint venture is a specialized manufacturer and distributor with convenient locations near the beautiful Taihu Lake, next to Jinghang Canal, only 2 km from an exit off Huning Expressway. The company occupies 11,898 m^2, of which 6,533 m^2 is leased land, and it employs 128 persons. All its bicycles are produced in accordance with Japanese Industrial Standards (JIS), and they have passed ISO 9001-2000 certification and Japanese SGS Group plant certification. Its main products are bicycles and electric bicycles, and 95% of its products are exported to Japan. It has a solid clientele and good reputation in the Japanese market. The company aims to meet client demands while keeping an eye on the international market and latest trends and recommending new products. It applies its scientific knowledge and standards, precision, and experience to all materials, structures, styles, colors, parts, and entire products to ensure that every bicycle meets international standards.

4.2.2.6. Changzhou Deyi Mechanical and Electrical Making Co., Ltd.

This company is located in the south Jiangsu Economic Development Zone, Jiangsu Province, at the Hengshan entry of the Shanghai-Nanjing Expressway. It has an area of more than 20,000 m^2. The company develops and produces electric bicycles and electric wheel hub motors. For 5 years, its products have been checked and thoroughly qualified through spot-checking at the provincial level. With an advanced product line of E-bikes and advanced production and inspection equipment, such as a meter that measures motor power, all the company processes are ISO 9001:2000 certified. With a production capability of 150,000 E-bikes and 600,000 sets of electric wheel hub motors, the company is poised to provide novel first-class products, brands, and service.

4.2.2.7. Changzhou Huajia Vehicle Industry Co., Ltd.

This company, along with Changzhou Xiaohe Huajia Telecommunication Lamp Factory, is located by the southern bank of the Yangtze Changjiang River in the Changzhou State High- and New-Tech Industrial Development Zone–Xiaohe Town, which is famous for automobile and motorbike fittings. Since its establishment, the company has been developing and manufacturing motorbike and electric bike lamps, plastic parts, and punched parts, as well as electric bikes themselves. The leading products — Yejie brand electric bike frames and plastic parts — have a good reputation in the Chinese electric bike industry. The company is 180 km from Shanghai to the east and 100 km from Nanjing to the west. It is within a 10-min car ride the Changzhou Airport, Luoshuwan entrance to the Shanghai-Nanjing Expressway, and Changzhou Harbor, and State Highways 101 and 218 connect and pass through the town. The company has a strong technical capability, advanced technologies, complete testing apparatus, various plastic injection machines, and a number of staff who are experienced in manufacturing vehicle lamps, plastic and punched parts, and vehicle frames. All of its products passed ISO 9001-2000 certification in 2004. Customer satisfaction is an important goal.

4.2.2.8. Changzhou Yufeng Vehicle Factory

Established in 1995, this factory is one of the largest manufacturers of electric tricycles in China, covering more than 3,800 m^2 and employing more than 30 technologists. Its main products are different electric leisure vehicles for senior citizens, tricycles for sightseeing, small electric tricycles, cargo electric tricycles, transformers, motors, high-intensity synthetic bodies, metal car bodies, glass-fiber-reinforced plastic car bodies, and plastic car parts. The factory employs excellent equipment, advanced techniques, strict quality management, and a complete set of checkout methods. It is a base for producing electric tricycles and parts that require advanced welding, spray painting, and drying equipment. Its highly qualified technologists and managers have abundant knowledge and practical experience. Yufeng brand electric tricycles are sold in China, Europe, the United States, and the Middle East. Special motors for these tricycles have passed national China Compulsory Certification (CCC) and are protected with liability insurance from the People's Insurance Co. of China (now PICC Holding). Many high-quality Yufeng electric tricycles have been sold; consumes have selected them as the "Name Brand of China" and an "Excellent Enterprise in China," and they have received many national patents.

4.3. Advantages of and Opportunites for Electric Two-Wheelers

In Beijing, roads are so crowded that the average speed attained by public transport vehicles is only 12 km/h. The average speed of a bike is 15 km/h and that of an E-bike is 20 km/h. Since E-bikes also do not occupy as much space on the road as cars, they could have an advantage in Beijing's public transportation system, perhaps playing an important role.

It has been stated that the development of cars could solve China's transportation problems. But in Shanghai and Beijing, the average speed at which a car can travel is only 20–30 km/h. This state of affairs leaves room to develop 20–35-km/h ETWs to reduce the crowd on the roads in these cities.

In the countryside of China, ETWs are very welcome. And although gas-fueled motorcycles are forbidden in big cities in China, they also are popular in the country. Because ETWs, when compared with gas-fueled motorcycles, are both more convenient (since drivers do not need to stop at gas stations) and safer (they travel at a slower speed), the opportunity for using them more in the Chinese countryside is greater.

4.4. Technology, Economic, and Policy Factors

E-bikes developed very rapidly before 2007. In 2007, however, the market for E-bikes was much worse than predicted. Some small E-bike companies stopped production or went out of business, although big companies were still increasing production. Some reasons for the slowdown in growth are as follows:

- Production capacity expanded too fast; production was much higher than demand.
- Too many new companies came into the market, causing stiff competition.

- The quality of the E-bikes was not stable. Quality was low on the basis of customer sampling.
- Consumers complained about bad quality and after-sale service.
- E-bike models were replaced so frequently (an average of two to three models were withdrawn from the market each month) that no parts were available after a model had been used for only a short while.
- The higher cost of raw materials (especially the much higher price of lead acid batteries) caused the price of E-bikes to increase.
- There was no national policy and there were no regional policies on E-bike use, and standards for E-bike use were still being debated.
- In the second half of 2006, some cities (like Zhuhai and Dongguan in Guangzhou Province) forbade electric bicycles to be driven on the road because of safety, management (the right of various vehicles to the road), and environmental (lead acid battery) issues. Some other local governments were also not encouraging their use.

In 2008, the ETW industry in China was confronted with two situations. On one hand, there was an unavoidable trend to protect the environment, save energy, and reduce emissions. The use of alternative energy became a very urgent concern, especially as oil prices rapidly increased. Hence, the use of an energy-saving electric vehicle should have been very welcome. On the other hand, however, the E-bike industry was plagued with embarrassingly vague standards, unreasonable competition, and high prices for raw materials, and consumers were becoming more aware of the problems as well as the benefits associated with their use. Apparently, China's E-bike industry would require reorganization, mergers, and closures in order to work toward the goal of selling 25 million E-bikes per year.

In the EU, however, imports of lead acid-battery E-bikes have been limited. The EU has advocated the use of Li-ion power batteries for E-bikes. Since 2007, there has been no tax return for exporting lead acid batteries. However, there has been a 17% tax return for exporting Li-ion batteries, which enhances their development.

It was suggested that development of China's EVs should go from small to big vehicles (from electric bicycles to motorcycles to plug-in electric cars to vans to buses to trucks). It was also suggested that gas-fueled motorcycles should be forbidden, and that new standards and ways of managing vehicle transportation (e.g., allowing special rights-of-way for E-bikes) should be implemented to encourage development of ETWs, especially those with Li-ion power batteries.

5. Chinese Government Policies

5.1. Government Plan

The Chinese government is formulating a policy for ETWs. The policy standards will be united yet address different vehicle specifications. Solving the ETW right-of-way problem will be a critical function of this policy.

The government has been financially supporting R&D on Li-ion power batteries. The development of this type of battery system is part of the "Energy Saving and New Energy

Vehicles" project, which is under the "Modern Transportation Technology Area" of the 863 Plan. Table 5-1 shows the project targets. The government is providing 66 million ¥ in funds (2008–2010) toward developing Li-ion power battery systems. The fund should be matched by industry at a scale of 1:1.

Table 5-1. Targets for Li-Ion Power Batteries under the Key Project, "Energy Saving and New Energy Vehicles," in the Modern Transportation Technology Area of the 863 Plan.

Parameter	Capacity (A·h)		
	8, 20	50	100
Power density (W/kg)	≥1800	≥700	≥500
Energy density (Wh/kg)			
LiFePO$_4$	≥65	110	≥110
LiMn$_2$O$_4$	≥70	≥120	≥120
Maximum discharging rate (C)	30 for 20 s	6 for 30 s	5 for 30 s
Maximum charging rate (C)	10 for 10 s	4 for 60 s	4 for 60 s
Unit cell impedance (mΩ)	≤2.0	≤3.0	≤2.5
Unit voltage bias (V)	≤0.02		
Unit cell capacity bias (%)	≤2		
Operating temperature (°C)	−25 to 60		
Storage temperature (°C)	−40 to 80		
Capacity retained in storage at room temperature for 28 days (%)	≥90		
State of charge estimation error (%)	≤5		
Safety	Pass the standard or specification		
Life of battery pack (km)	150,000 (LiFePO$_4$), 100,000 (LiMn$_2$O$_4$)		
Reliability of battery pack	Operated normally; could run 30,000 km		
Cost (¥/Wh)	≤3	≤2	≤2

5.2. Relevant Standards

New national standards for ETWs will come out soon. Definitions for Chinese ETWs are suggested here in order to better manage them.

Electric bicycles
Speed: ≤20 km/h
Vehicle weight: ≤48 kg
Motor power: ≤240 W
Battery pack voltage: ≤48 V

Light electric motorcycles
Speed: ≤40 km/h
Vehicle weight: 55–80 kg
Motor power: ≤500 W

Battery pack voltage: ≤48 V
Only for one-person ride

Electric motorcycles
Speed: ≤80 km/h
Vehicle weight: 80–150 kg
Motor power: 2500 W
Battery pack voltage: ≤48 V

5.3. Income Tax Incentives

In 2007, the return on taxes for exporting lead acid batteries was cancelled. However, there is still a 17% return on taxes for exporting Li-ion batteries, which enhances their development.

Here is an example of some income tax incentives and how they apply to an example enterprise called ABC. According to applicable PRC income tax laws and regulations:

- Both an enterprise located in Shenzhen and the district in which its operations are located are subject to a 15% enterprise income tax.
- Foreign-invested manufacturing enterprises, starting from their first profitable year, are entitled to a two-year exemption from the enterprise income tax, followed by a three-year 50% reduction in the enterprise income tax.
- PRC companies, starting from their first profitable year, are entitled to a two-year exemption from the enterprise income tax, followed by a three-year 7.5% reduction in the enterprise income tax.

Being a PRC company, ABC enterprise is exempted from any income tax for the first two years. For the following three years, it is subject to a reduced income tax rate of 7.5%.

Some preferential tax treatment is also applicable to ABC, and it is fully exempt from any income tax during a tax holiday. (A tax holiday is a designated period — the month of June each year — during which companies do not pay income tax on equipment purchases or any other incurred business expenses.) Also, because of the additional capital invested in ABC, it was granted a reduced income tax rate of 1.7% for two years.

Finally, in order to encourage the investors to introduce advanced technologies in China, the PRC also offers additional tax incentives to enterprises that are classified as a foreign-invested enterprises with advanced technologies. If an enterprise qualifies for this designation, then it pays 1.7% in taxes for an additional three years. It can then renew this status and continue to pay a reduced income tax. As long as ABC maintains this designation, it may apply to the tax authority to extend its current reduced tax rate of 1.7% for another three years.

6. Future Challenges and Recommendations

Li-ion batteries offer very high power while charging and discharging. Further improvements — such as power at low temperatures — might also be possible. The main challenge with regard to this technology, in addition to reducing its cost, is attaining an acceptable operating life, particularly at 40°C. Battery manufacturing companies and R&D organizations worldwide are now making major efforts to mitigate the relatively rapid fading of the $LiMn_2O_4$ Li-ion battery that occurs at elevated temperatures. The degree of improvement that will be achieved is, however, difficult to anticipate.

The basic chemistry and design of Li-ion ETW cells are quite similar to those of small consumer cells, which suggests that the basic manufacturing processes for ETW and EV batteries should be well understood. The manufacture of Li-ion cells is known to require a higher level of process control and precision than most other types of battery manufacturing, and, as a result, scrap rates tend to be higher. Most, if not all, producers of small Li-ion batteries have experienced product recalls or production shutdowns as result of reliability issues or safety incidents. Extrapolating this experience to the much larger HEV cell with its thinner electrodes indicates that scaling up the production of HEV cells from the current early pilot level will probably be slow and costly. If Li-ion HEV batteries are to become commercially viable, issues associated with their operating life and tolerance to abuse will need to be resolved first, and then the unit cost of the technology will need to be reduced, at least to the levels projected for Ni-MH batteries. These steps will help to reduce the cost and improve the performance of Li-ion battery technology for ETWs and EVs.

Two recommendations are made. First, a DOE program official(s), along with experts, should visit China to study the battery technology industry firsthand and make arrangements for benchmarking Chinese battery technology in the United States. Chinese companies have expressed a strong interest in making battery technology available for benchmarking. The timing is right, and interest in working with the United States is very strong. Second, DOE and the Chinese Government (i.e., the Ministry of Science and Technology) should work together to set up a battery workshop in China and invite U.S. and international companies to participate. This effort will help the Li-ion battery and ETW industries work with their counterparts to more rapidly develop advanced, reliable, low-cost Li-ion batteries.

Bibliography

7.1. Electric Bike Web Sites

[1] http://www.usatoday.com/money/autos/environment/
[2] http://www.evehicle.com.au/
[3] http://www.electrikmotion.com/
[4] http://www.50cycles.com/
[5] http://www.optibike.com/
[6] http://www.treehugger.com/files/2006/10/schwinns_new_line_of_electric_bikes.php
[7] http://www.electricbicyclesusa.com/
[8] http://www.electricbikesales.co.uk/

[9] http://www.electricvehiclesnw.com/
[10] http://www.electrikmotion.com/
[11] http://www.electric-bikes.com/
[12] http://www.globalsources.com/manufacturers/Electric-Bike-3-Wheel.html
[13] http://www.b2bfreezone.com/product-search/electric-three-wheeler.htm
[14] http://www.electric-bikes.com/aboutus/index.html
[15] http://www.financialexpress.com/news/Electric-bikes-and-3wheelers-to-hit-Punjab-roadssoon/294956/
[16] http://www.allbusiness.com/transportation
[17] http://www.indiaautomotive.net/2008/04/electrotherm-plans-more-powerful.html
[18] http://www.ultramotor.com/userfiles/News_24092007_443_59.htm
[19] http://infospice.blogspot.com/2008/02/nano-puts-brakes-on-electrotherm-3.html
[20] http://www.livemint.com/2008/07/24220523/Scooters-India
[21] http://express-press-release.net/3 1 /STANDARD%20LAUNCHES%20ECO-FRIENDLY%20BATTERY%20OPERATED%20ELECTRIC%20-%20BIKE%20IN%20INDIA.php
[22] http://www.thehindubusinessline.com/2006/11/22/stories/2006112204310300.htm
[23] http://nextbigfuture.com/2007/08/clean-vehicles
[24] http://www.vicky.in/straightfrmtheheart/preview-on-tvs-scooty-teenz-electric-and-threewheeler/
[25] http://www.ec21.com/ec-market/three_wheelers.html
[26] http://dir.indiamart.com/impcat/electric-bike.html
[27] http://www.surfindia.com/automotive/two-wheelers/
[28] http://www.cleanairnet.org/caiasia/1412/propertyvalue-14302.html
[29] http://ieeexplore.ieee.org/xpl/freeabs_all.jsp?tp=&arnumber=4156558&isnumber=4156546
[30] http://www.dancewithshadows.com/autoindia/electric-scooters-in-india-upcoming-launches-and-existing-scooters/
[31] http://www.chinasuppliers.globalsources.com/china-suppliers
[32] http://www.electric-bikes.com/intro.html
[33] http://www.directdiscountusa.com/?gclid=CNvtvZPsx5YCFQZlswodCHuvyg
[34] http://energy
[35] http://nariphaltan.virtualave.net/att42.htm
[36] http://nariphaltan.virtualave.net/itm00009.htm
[37] http://www.aboutmyplanet.com/black-gold/electric-rickshaws-transportation
[38] http://findarticles.com/p/articles/mi_qa3650/is_199703/ai_n8738529
[39] http://www.innovationsofindia.com/list_of_innovations/elecsha.htm
[40] http://www.flyingpigeone-bike.com/en/about.asp
[41] http://www.ecvv.com/manufacturers/P1CM0V0IY303CN/Electric-Bicycle.html
[42] https://www.up.ac.za/dspace/bitstream/2263/6050/1/021.pdf
[43] http://www.szzmcyh.com/en/about.asp
[44] http://www.bikeforums.net/forumdisplay.php?f=25 8
[45] http://www.szchuantian.com/en/about.asp
[46] http://www.electricvehiclesnw.com/main/ebike-comp.htm
[47] http://www.atob.org.uk/Electric_price_tag.html
[48] http://www.myebike.com/advantage/Anelectricbikeasanalternativetoacar.html

[49] http://pubsindex.trb.org/document/view/default.asp?lbid=848390
[50] http://www.ebikes.ca/sustainability
[51] http://www.wheelsunplugged.com/show_story_new.aspx?sid=4&iid=7
[52] http://www.the-infoshop.com/study/go9614_bicycles_toc.html
[53] http://www.cleanairnet.org/lac/1471/articles-40938_resource_1.pdf
[54] http://www.biketaiwan.com/new/script/Newsletter/news_main.asp#Special%20Reports_1
[55] http://www.adb.org/Documents/Reports/Energy-Efficiency-Transport/chap02.pdf

7.2. Other Sources

[1] Anderman, M. (2003). "Brief Assessment of Improvements in EV Battery Technology Since the BTAP June 2000 Report," California Air Resources Board, Sacramento, Calif.

[2] Bos, I., van der Heijden, R., Molin, E. & Timmermans, H. (2004). "The Choice of Park & Ride Facilities: An Analysis Using a Context-Dependent Hierarchical Choice Experiment," in *Proceedings of 83rd Annual Meeting of the Transportation Research Board*, Jan., pp. 2–3.

[3] Bouwman, M. E. (2000). *An Environmental Assessment of the Bicycle and Other Transport Systems,* University of Groningen, Netherlands, http://www.velomondial.net/velomondiall2000/ PDF/BOUWMAN.PDF.

[4] Cherry, C. (2005). *China's Urban Transportation System: Issues and Policies Facing Cities,* UCB-ITS-VWP-2005-4, University of California at Berkeley.

[5] Cherry, C. (2006). "Implications of Electric Bicycle Use in China: Analysis of Costs and Benefits," presented at Volvo Summer Workshop, Center for Future Urban Transport, University of California at Berkeley, July 24–26.

[6] Cherry, C. & Cervero, R. (2006). *Use Characteristics and Mode Choice Behavior of Electric Bikes in China,* UCB-ITS-VWP-2006-5, University of California at Berkeley.

[7] Cherry, C., Weinert, J. & Ma, Z. (2006). *The Environmental Impacts of Electric Bikes in China,* Working Paper 2006, Center for Future Urban Transport, University of California at Berkeley.

[8] Chiang, W. L. (1996). "Electric Scooters in Chinese Taipei," presented at APEC Transportation Forum on Energy, Technology and the Environment, Auckland, New Zealand, Apr.

[9] *China People's Daily Online.* (2006). "China Expected to Get Back on Their (Electric) Bicycles," May, http://www.chinadaily.com.cn/cndy/2006-05/16/content_590600.htm (accessed Nov. 2006).

[10] Chiu, Y. C. & Tzeng, G. H. (1999). "The Market Acceptance of Electric Motorcycles in Taiwan Experience through a Stated Preference Analysis," *Transportation Research Part D,* Vol. 4.

[11] Dell, R. M. & Rand, D. A. J. (2001). *Understanding Batteries,* The Royal Society of Chemistry, Cambridge, United Kingdom.

[12] Energy Foundation China. (2005). *Electric Utilities,* http://www.efchina.org/programs.electricutilities.cfm (accessed Dec. 12, 2005).

[13] *General Technical Standards of E-Bike (GB1 7761-1999).* (1999). Chinese Federal Government, Beijing, PRC.
[14] Government of the ROC. (1995). *Current Situation of Motorcycles Pollution Control in Republic of China,* Environmental Protection Administration, June.
[15] Gu, Q. & Gu, S. H. (2001). "Preferred Green Transportation Tools: Bicycle," *Journal of Economy & Management,* No. 4, p. 35.
[16] *Guangzhou Daily.* (2006). "Guangzhou Bans Electric Bikes," Nov.
[17] Jamerson, F. & Benjamin, E. (2005). *Electric Bicycle World Report,* 7th edition with 2005 update, International Energy Agency, Paris.
[18] Jang, H. (2006). "Electric Vehicles Status in China," in *Proceedings of Light Electric Vehicle International Conference,* pp. 223–237.
[19] Jet, P. & Shu, H. (1996a). "Electric 'Scooters' in Taiwan, ROC," presented at 13th International Electric Vehicle Symposium, Osaka Japan, Oct.
[20] Jet, P. & Shu, H. (1996b). "The Development of Electric Motorcycles in Taiwan, ROC," presented at ROC Electric Vehicles International Symposium, Taiwan, Feb.
[21] Jet, P., Shu, H., Chiang, W. L., Lin, B. M. & Cheng, M. C. (1996). "The Development of the Electric Propulsion System for the Zero Emission Scooter in Taiwan," presented at Small Engine Technology Conference, Yokohama, Japan, Oct.
[22] Jet, P., Shu, H., Chiang, W. L., Lin, B. M. & Cheng, M. C. (1998). "The Development of the Electric Propulsion System for the ZES2000 in Taiwan," presented at the 15th International Electric Vehicle Symposium, Brussels, Belgium, Oct.
[23] Kang, L. (2004). *Feasibility Study on Introducing Fuel Cell Two-Wheeler Technologies into Shanghai Market,* Shanghai Jiaotong University Report, Mar., in Chinese, http://www.sj998. com/subject/show_content.asp?id=590 (accessed June 2006).
[24] Lin, B. M. (2006). "The Promotion of Electric Vehicles in Taiwan," presented at 2006 Taipei Power Forum, Session I-EV, Taipei, Taiwan.
[25] Litman, T. (1999). "Quantifying the Benefits of Non-motorized Transport for Achieving TDM Objectives," VTPI, http://www.vtpi.org/nmt-tdm.pdf (accessed Nov. 2004).
[26] Liu, X. M. & Chen, J. C. (1998). "The Sustainable Developing Strategy of Urban Transportation in China," presented at Beijing Transport and Environmental Protection Seminar, Beijing, PRC, May, pp. 192–217.
[27] Lu, X. & Chen, X. (1996). *Passenger Transport Planning and Urban Development,* Press of Huadong Polytechnic University, Tinjin, PRC.
[28] Mechanical Industry Research Laboratory. (1995). *Electric Motorcycle Technology Development,* Tokyo, Japan, July.
[29] National Bureau of Statistics, 2004, *List of the Average Number of Consumer Durables of Urban Dwellers,* Beijing, China, http://2 10.72. 32.26/yearbook200 1 /indexC.htm (accessed June 2004).
[30] National Bureau of Statistics. (2005). *China Statistical Yearbook,* Beijing, China. National Bureau of Statistics, 2007, *China Statistical Yearbook,* Beijing, China.
[31] Neupert, H. (2007). "Proposal to Accelerate Market Size and Quality of E-Bikes/Pedelecs in Europe," presented at Light Electric Vehicle Conference, Taipei, Taiwan. July.
[32] Ni, J. (2004). *Feasibility Study: Proposal for the Manufacture of Mini Electric Cars Based on Experience with Large Scale Manufacture of Light Electric Vehicles,* China

LEV Development and Strategic Study Report, Luyuan Bicycle Company, p. 49, in Chinese.

[33] Origuchi, M., et al. (1996). "Development of a Lithium-Ion Battery System for EVS," SAE Paper 970238, presented at Electric Vehicles Symposium, Osaka, Japan.

[34] Patil, P. (2008). *Developments in Lithium-Ion Battery Technology in the Peoples' Republic of China,* ANL/ESD/08- 1, Argonne National Laboratory, Lemont, Ill., Jan.

[35] Pek, J. (2002). "China's Tailpipe Tally: The World's Biggest Nation 'Modernizes' with More Cars," *China Vehicle Monthly,* May, http://www.emagazine.com/novemberdecember/2002/ 1 102curr _china.html (accessed June 2004).

[36] *People's Daily Online.* (2005). "Traffic Accidents Impair China's GDP Growth," Dec., english.people.com.cn/200512/16/eng20051216_228514.html (accessed July 2006).

[37] Rossinot, E., Lefrou, C.& Cun, J. P. (2003). A Study of the Scattering of Valve Regulated Lead Acid Battery Characteristics," *Journal of Power Sources, 114(1)*, 160–169.

[38] Shah, J. & Harshadeep, N. (2001). "Urban Pollution from Two Stroke Engine Vehicles in Asia: Technical and Policy Options," presented at Regional Workshop on Reduction of Emissions from 2-3 Wheelers, Hanoi, Vietnam, Sept. 5–7, http://www.adb.org/Documents/Events/2001/ RETA5937/Hanoi/default.asp?p=vhclemsn&pg=workshop.

[39] Suzuki, K. (2007). "Advanced Technologies Driving Pedelec Market and Excellent Performance and Safety by Li-Ion Battery for Pedelec," presented at Light Electric Vehicle Conference, Taipei, Taiwan, Aug.

[40] Taiwan EPA (Environmental Protection Agency). (1998). "Taiwan Steps up Promotion of Electric Motorcycles," *Environmental Policy Monthly,* Vol. 2.

[41] TVMA (Transportation Vehicle Manufacturers Association) of Taiwan. (1996a). *Transportation Vehicle Manufacturers Association Monthly Report,* Jan.

[42] TVMA of Taiwan. (1996b). "The Status of Electric Motorcycle Development," May.

[43] Weinert, J., Ma, Z. & Cherry, C. (2006). "The Transition to Electric Bikes in China: History and Key Factors for Rapid Growth," in *Proceedings of the 22nd Electric Vehicle Symposium Conference,* Yokohama, Japan, Oct. 25–2 8.

[44] Wilson, D. G. (2004). *Bicycling Science,* The MIT Press, Boston, Mass.

[45] Wu, D. (2007). "Lithium-Ion Phosphate Batteries: Enviro-friendly Technology Offered by Pihsiang Energy Technologies," presented at Light Electric Vehicle Conference: Battery Safety, Taipei, Taiwan, Aug.

[46] *Xinhua Net.* (2006). "Guangzhou Bans Electric Bikes," Nov., in Chinese, http://news.xinhuanet. com/fortune/2006-1 1/03/content _5284544.htm (accessed Nov. 2006).

[47] Xu, J. Q., Zhang, Y. D. & Mei, B. (1995). "Bicycle Travel Characteristics and Proper Travel Distance," *Urban Transport,* Vol. 2, p. 30, in Chinese.

[48] Yang, M. H. (2007). "Outlook of Future Li-Ion Battery Chemistries for Safety Improvements," presented at Light Electric Vehicle Conference: Battery Safety, Taipei, Taiwan, Aug.

[49] Yang, M. H., Lin, B. M. & Chang, H. S. (2005). "LEV Progress in Taiwan," presented at Electric Vehicles Symposium 21, Paris, France.

[50] Zhang, Y. (2000). "Bicycle Express Delivery — A New Scenery of City Service," *Communication Management*, No. 7, p. 2.

[51] Zhang, Y., et al. (2005). "Research on the Development of Bicycle Traffic in Big Cities," *Journal of Transportation Engineering and Information*, Vol. 3, No. 4, Dec., in Chinese.

[52] Zheng, Z. (1998). *Modern Urban Transportation*, People Transportation Press, Beijing, PRC.

APPENDIX A. CHINESE EXPERTS INTERVIEWED ABOUT LITHIUM-ION BATTERIES

1. Dr. Wang Zhen-po, Professor
   Beijing Institute of Technology
   BTI EV Center of Engineering and Technology
   No. 5 South Zhonggancun South Street
   Haidian District
   Beijing, 100081

2. Dr. Wang Wenwei, Professor
   Beijing Institute of Technology
   BTI EV Center of Engineering and Technology
   No. 5 South Zhonggancun South Street
   Haidian District
   Beijing, 100081

3. Dr. He Hong-Wen, Professor
   Beijing Institute of Technology
   BTI EV Center of Engineering and Technology
   No. 5 South Zhonggancun South Street
   Haidian District
   Beijing, 100081

4. Meng Xiangfeng, Project Assistant
   Beijing Institute of Technology
   BTI EV Center of Engineering and Technology
   No. 5 South Zhonggancun South Street
   Haidian District
   Beijing, 100081

5. Wu Ningning, Vice Director
   Research Institute
   CITIC Guoan MGL
   MGL New Energy Technology Co., Ltd.
   18 Biafuquan Road
   Changping District
   Beijing, 102200

6. Dr. Qi Lu, Vice Chairman
   CITIC Guoan Group
   Guandondian North Street
   Chaoyany District
   Beijing, 100020

7. Dr. Tian Guangyu, Professor
   Tsinghua University
   Department of Automotive Engineering
   Beijing, 100084

8. Dr. Lin Chengtao, Professor
   Tsinghua University
   Department of Automotive Engineering
   Beijing, 100084

9. Wang Longzhang, Manager
   Corporate Development Department
   Aluminum Corporation of China Ltd.
   62 North Xizhimen Street
   Beijing, 100068

10. Deng Jie, Business Manager
    Project Division
    Science and Technology Department
    Aluminum Corporation of China Ltd.
    62 North Xizhimen Street
    Beijing, 100068

11. Zhang Jilong, Director
    Science and Technology Department
    Aluminum Corporation of China Ltd.
    62 North Xizhimen Street
    Beijing, 100068

12. Dr. Chen Jun, General Manager
    Wanxiang Group
    Wanxiang EV Company, Ltd.
    Hangzhou, China

13. Xuezhe Wei, Professor
    School of Automobile Engineering
    Tongji University
    Shanghai FCV Powertrain Co., Ltd.
    4800 Cao An Road
    Shanghai, 201804

14. Wei Yang
    School of Automobile Engineering
    Tongji University
    Shanghai FCV Powertrain Co., Ltd.
    4800 Cao An Road
    Shanghai, 201804

15. Sun Zechang, Vice-Director
    Automotive Engineering College
    Tongji University
    4800 Cao An Road
    Shanghai, 201804

16. Wang Jiayuan, Post Doc Student
    School of Automobile Engineering
    Tongji University
    Shanghai FCV Powertrain Co., Ltd.
    4800 Cao An Road
    Shanghai, 201804

17. Xu Wei, Postdoc Student
    School of Automobile Engineering
    Tongji University
    Shanghai FCV Powertrain Co., Ltd.
    4800 Cao An Road
    Shanghai, 201804

18. Dianna Dong, Graduate Student
    School of Automobile Engineering
    Tongji University
    Shanghai FCV Powertrain Co., Ltd.
    4800 Cao An Road
    Shanghai, 201804

19. Xuefeng Gao, Vice President
    DLG Battery (Shanghai) Company, Ltd.
    3492 Jinqian Road
    Qingcun Town
    Fengxian District
    Shanghai, 201406

20. Rita Chen, General Manager
    DLG Battery (Shanghai) Company, Ltd.
    3492 Jinqian Road
    Qingcun Town, Fengxian District
    Shanghai, 201406

21. Gao (King) Pengkun, Chief Engineer
    DLG Battery (Shanghai) Company, Ltd.
    3492 Jinqian Road
    Qingcun Town
    Fengxian District
    Shanghai, 201406

22. Stewart G. Graham, Director of Operations
    K2 Energy Solutions, Inc.
    1125 American Pacific Drive, Suite C
    Henderson, NV 89074

23. Phegn Chen, General Manager
    A-SI-KA Electric Bike Company, Ltd.
    155 Baochen Road, Room 702
    New min Du chen
    Shanghai, 201100

24. Wang Xiang Min, Director
    The Administrative Committee
    of New Guangming District
    Economic Development Office
    Administrative Bldg. of ACNGD, No. 1, Room 406
    Tangming Road, New Guangming District
    Shenzhen, 518108

25. Sandy Xinyu Wang, Executive Manager
    City University of Hong Kong
    Shenzhen Virtual University Park
    Room A307
    Shenzhen Hi-Tech Industrial Park
    Shenzhen, 518057

26. Huang Zheng Yao, R&D Manager
    Shenzhen Wisewod Technology Company, Ltd.
    C Spot, Lian tang Industrial Park
    Gongming, Bao'an District
    Shenzhen, 518106

27. Hu Ji Tao, R&D Engineer
    Shenzhen Wisewod Technology Company, Ltd.
    C Spot, Lian tang Industrial Park
    Gongming, Bao'an District
    Shenzhen, 518106

28. George Pan, Board Chairman and CEO
    Shenzhen HighPower Technology Company, Ltd.
    Luoshan Industrial Zone
    Pinghu, Longgang
    Shenzhen, Guangdong, 518111

29. Wallace Liu, Marketing Analysis Engineer
    Shenzhen HighPower Technology Company, Ltd.
    Luoshan Industrial Zone
    Pinghu, Longgang
    Shenzhen, Guangdong, 518111

30. Kevin Wen, Vice R&D Manager
    Shenzhen HighPower Technology Company, Ltd.
    Luoshan Industrial Zone
    Pinghu, Longgang
    Shenzhen, Guangdong, 518111

31. J. Simon Xue, Chief Technical Officer
    Shenzhen B&K Electronics Company, Ltd.
    B&K Lithium Ion Battery
    Hongfu Industrial Park
    Dalang, Huarong Road, Longhua
    Baoan District, Shenzhen

32. Zheng Rongpeng, Manager
    Shenzhen BAK Battery Company, Ltd.
    BAK Industrial Park
    Kuichong Street, Longgang District
    Shenzhen, 518119

33. Daotan Liu, R&D Engineer
    Shenzhen BAK Battery Company, Ltd.
    BAK Industrial Park
    Kuichong Street, Longgang District
    Shenzhen, 518119

34. Jason Li, Electrical Design Engineer
    Shenzhen BAK Battery Company, Ltd.
    BAK Industrial Park
    Kuichong Street, Longgang District
    Shenzhen, 518119

35. Dr. Qingyu Li, Professor
    College of Chemistry and Chemical Engineering
    Guangxi Normal University, Guangxi, 541004

36. Zeng Jian Yi, General Manager
    Shenzhen Herewin Technology Company, Ltd.
    Haohaihong Industrialized Country
    4th Industrial Park, Gonghe Village
    Shajing Town, Baoan District
    Shenzhen, 518104

37. Guo Yong Xing, Asst. General Manager
    Shenzhen Herewin Technology Company, Ltd.
    Haohaihong Industrialized Country
    4th Industrial Park, Gonghe Village
    Shajing Town, Baoan District
    Shenzhen, 518104

38. Yuanbin Lie, Chief Engineer
    Shenzhen Redway Battery Company, Ltd.
    Shenzhen, 518104

39. Zhou Zhi Cai, CEO
    Shenzhen BYN Battery Company, Ltd.
    3F, C1, NYF NO. 4 Bitou Industrial Zone
    Song Gang, Shenzhen City

40. Liu Wei-Ping, CTO
    BYD Company Ltd.
    No. 1, Baoping Road, Baolong, Longgang
    Shenzhen, 518116

41. Li Ke, Section Chief
    BYD Company Ltd.
    No. 1, Baoping Road, Baolong, Longgang
    Shenzhen, 518116

42. Dr. Qilu, Professor
    Department of Applied Chemistry
    College of Chemistry and Molecular Engineering
    Peking University
    No. 3 Zhonggguanncun Bel er Tiao Street
    Haidian District
    Beijing, 100080

43. Jinhua Zhang, Vice President
    China Automotive Technology and Research Center
    Block 7, Phase II
    186 Western Road, 4th South Ring Road
    Fengtai District
    Beijing, 100070

44. Hou Fushen, Director
    Hi-Tech Development Department
    China Automotive Technology and Research Center
    F/9 Block 7, Phase II
    188 Western Road, 4th South Ring Road
    Fengtai District
    Beijing, 100070

45. Guan Yu, Engineer
    China Automotive Technology and Research Center
    F/9 Block 7, Phase II
    188 Western Road, 4th South Ring Road
    Fengtai District
    Beijing, 100070

APPENDIX B. PRESENTATION ON LITHIUM-ION BATTERY TECHNOLOGY

B.1. Current Status of HEV Batteries

- Conventional lithium-ion batteries for HEVs appear to be almost ready for commercialization.
- The major focus is still on reducing their cost.
- Concerns regarding their performance at low temperatures and their ability to tolerate abuse still remain. However, emerging technologies with nanostructure materials ($Li_4Ti_5O_{12}$ or $LiFePO_4$) appear to address these concerns.
- Batteries, even those incorporating "stable" materials, require appropriate thermal management controls and electronic protection circuits to extend battery life and avoid thermal runaway.
- The battery-life projections of 10–15 years are based on limited data.

B.2. Lithium-Ion Battery Technology

- Advantages
  - Highest energy storage
  - Lightweight
  - No memory effect
  - Good cycle life
  - High energy efficiency
  - High unit cell voltage
- Disadvantages
  - Relatively expensive
  - Electronic protection circuitry

- Thermal runaway concern
- 3-h charge
- Not tolerant of overcharge

B.3. Microsun's High-Power Lithium-Ion Battery

- Cell specifications
  - Commercially available Type HPPC 18650 (high power)
  - Cutoff voltage: 3.0 to 4.2 V
  - Cell rated capacity: 1.6 A·h
  - Maximum discharge: 20 A
  - Maximum charge: 5 A (80% in <15 min)
  - Cycle life: >1,000 cycles (80% charge)
- Module specifications
  - Module design: 4 series × 5 parallel
  - Nominal voltage: 14.4 V
  - Module capacity: 8 A·h
  - Maximum discharge: 50 A continuous, 100 A of 10-s pulses
  - Charge time: <1 h (80%)
  - Fully integrated cell balancing, safety circuit, and thermal management
  - Dimensions: 4.30 in. high × 5.30 in. wide × 2.75 in. long

B.4. Gold Peak Industries North America

Battery is standard cell Type 18650 at 2,650 mAh, 42.5 g, 230 Wh/kg, 0.67 A/cell (three parallel). Data on both versions of Gold Peak's high-capacity batteries are shown in the table below.

Parameter	LiSO$_2$ Primary	Lithium-Ion
Current (A)	20	67
Capacity (A·h)	8.25	2.65
Voltage (V)	2.72	3.67
No. of cells	1	1
Energy (Wh)	22.44	9.73
Weight (kg)	0.0850	0.042 1
Power (W)	264	231

B.5. 26650 Lithium-Ion Battery Manufactured by A123

- Capacity: 2.3 A·h
- Energy: 7.6 Wh (110 Wh/kg)
- Nominal voltage: 3.3 V
- Cylindrical cell dimensions: 25.9-mm diameter, 65.4-mm high
- Cell volume: 34.45 cm^3

- Cell mass (without external tabs): 70 g
- Impedance (1 kHz): 8 MΩ
- Impedance (10 A, 10 s): 15 MΩ
- Operating temperature range: −30 to +60°C

B.6. Kokam America

- Fast charge capability: maximum 3°C
- High discharge capability: 10 to ~20°C
- High power density: >1,800 W/kg (high-power cell)
- Longer cycle life: >2,500 cycles at 80% depth of discharge (DOD)
- Wide operating temperature range: −30 to about +60°C
- Environmentally friendly: zero emissions
- Low energy consumption: lightweight
- Maintenance-free operation
- Low heat emission in high-discharging mode

Parameter	Power Cell	Energy Cell
Energy density		
Wh/kg	120	200
Wh/l	240	400
Power density		
W/kg	2,400	550
W/l	4,800	900

B.7. Electro Energy; Mobile Products, Inc.; and Bi-Polar Lithium-Ion Battery Technology

- $LiCoO_2$ chemistry
- Cell capacity: 20 A·h
- Cell is capable of 5°C charge and discharge
- 8 stacks of cells, total number of cells is 112
- Total capacity: 160 A·h
- System energy: 8 kWh
- Cell weight: 100 lb
- Additional battery hardware weight (5–10 lb) required

B.8. Saft High-Power Lithium-Ion Cells (Vl20P)

- Nominal voltage: 3.6 V
- Average capacity: 20 A·h, 1 C after charge to 4.0 V/cell
- Minimum capacity: 18.5 A·h, 1 C after charge to 4.0 V/cell
- Specific energy: 187 Wh/kg

- Specific power: 1,811 W/kg
- Cell dimensions: 41-mm diameter, 145-mm high
- Typical cell weight: 0.8 kg

B.9. High-Power Toyota 12-A·H–Cell Lithium-Ion Battery

- Voltage: 3.6 V
- Capacity: 12 A·h
- Specific power: 2,250 W/kg
- Specific energy: 74 Wh/kg
- Weight: 580 g
- Dimensions: 120-mm long, 25-mm wide, 120-mm high

B.10. Current Status of Lithium-Ion Technology

Conventional lithium-ion technology
- Accurate SOC
- Excellent power density
- Good energy density
- *Well matched for charge-sustaining*

Emergent lithium ion technology (titanate anode or iron phosphate cathode)
- SOC determination problematic
- Good power density
- Very good energy density
- *Well matched for PHEVs and potentially charge-sustaining HEVs*

NiMH
- Difficult to ascertain the SOC accurately
- Good power density
- Abuse tolerant and proven technology
- Moderate energy density; *good for charge-sustaining HEVs*

	HEV (Cell Energy, 10-s Power)	**EV** (Cell Energy, 30-s Power at 80% DOD)
Conventional Li-ion	70 Wh/kg 2,500 W/kg	140 Wh/kg 500 W/kg
Emergent Li-ion		100 Wh/kg* 2,000 W/kg
NiMH	50 Wh/kg 1,000 W/kg	65 Wh/kg 150 W/kg

* Projected

B.11. Challenges in Developing Large Lithium-Ion Batteries

- Abuse tolerance: material and battery management
- Cost: cathode selection, volume, standardization, packaging, battery management
- Life: cathode selection, operating temperature, packaging
- Performance in extreme temperatures: all aspects of chemistry

B.12. Status of Lithium-Ion Batteries versus Goals for Power-Assist Hevs

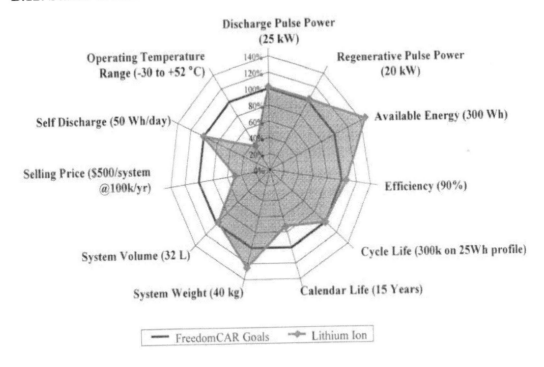

B.13. HEV and PHEV Battery Requirements (Module Basis)

Parameter	HEV	PHEV-20	PHEV-60
Zero-emissions vehicle (ZEV) range (mi)	0	20	60
Battery capacity (kWh)	<3	6	18
Cell size (range corresponds to battery voltage of 400–200 V)	5–10	15–30	45–90
Specific energy (Wh/kg)	>30	~50	~70
Specific power (W/kg)	~1,000	~440	~390
Cycle life			
Deep (80% depth of discharge)	Not available	>2,500	>1,500
Shallow (±100 Wh)	200,000	200,000	200,000

B.14. Requirements for Batteries for Transportation

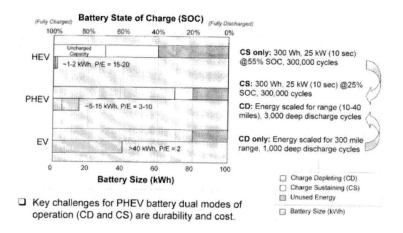

- Key challenges for PHEV battery dual modes of operation (CD and CS) are durability and cost.

B.15. Status of Advanced Battery Technology Development

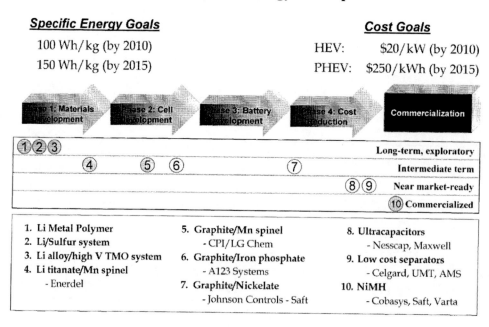

APPENDIX C. CHINA'S ELECTRIC BIKE INDUSTRY AND ETW MODELS PRODUCED

This is a typical list of companies that make ETWs. You can get more information from the Internet.

Electric Bicycle JP-ES001

****BEST SERVICE+BEST QUALITY+BEST PRICE=BEST COOPERATION****
FOR ANY INQUIRY,MAIL ME: Model No.:JP-ES001 Product ...
[Keywords: electric scooter, electric bike, bicycle]
See All Items(4) from Yongkang Jinpeng Hardware Manufacture Co.,Ltd
[China] TRADE PRO

Electric Bicycles with Li-ion Battery (JSL-TDH038XE)

Model Number:JSL-TDH038XE 1).Motor:250W(220W) High speed brushless DC hub intelligent motor 2).Battery:24V,8AH ...
[Keywords: electric bicycles, electric bikes, Electric bike]
See All Items(17) from Wuyi J.S.L Hardware Machinery Co., Ltd. [China]
TRADE PRO

Electric Bicycle with Light Weight & Simple Design(WZEB1834)

Electric Bicycle with Light Weight & Simple Design WZEB1834 1. User age: > 16 2. Motor: 180-250W brushed 3. ...
[Keywords: Electric Bicycle, Electric Bike, Bike]
Wiztem Industry Co., Ltd. [China] TRADE OK

Folding Electric Bike,Mini Bike,Electric Bicycle (YM-EB006)

Frame : Steel Tyres : 16 ? Rim : Aluminum Brakes(front) : ?V ?brakes Front fork : Shock Absorber Handle ...
[Keywords: electric folding bicycle, aluminium electric bicycle, shoes]
China Weiyi Industry & Trade.,Co Ltd [China] TRADE OK

Electric Bicycle

type:TDR552Z mptor:180W high speed motor battery:24V, 4.2Ah break:hand break wheel size:18*1.75 seat ...
[Keywords: Electric Bicycle, bicycles]
See All Items(2) from Jiangsu Hongdou Imp&Exp Co.Ltd [China]

Electric Bicycle

1. Dimensions:1680*650*1080(L x W x H mm) 2. Wheelbase: 1240(mm) 3. Gross Weight:100(kg) 4. Net Weight:90(kg) 5. ...
[Keywords: electric bicycle]
See All Items(15) from Wuxi Xufeng Electric Bicycle Co.,Ltd [China]

Electric Bicycle TDRX009TZ

Description Li RATED VOLTAGE DC36V WHEEL DIA 16" POWER 250W MAX SPEED 25Km/h RANGE 50-60Km WEIGHT(WITH ...
[Keywords: electric bike, electric bicycle, electric scooter]
Jinan Allied Inte Trade Co.,Ltd. [China]

Electric Bicycle

YY01E Technical specification Motor: 250W Battery: 36V12AH Low-voltage protection value: 31.5+0.3V Over-carrent ...
[Keywords: bicycles, electric bicycles, electric scooter]
See All Items(2) from Zhejiang Yongyuan Trading Corperation [China]

Electric Bicycle (CEB02)

1)power of motor:200W-350W brushless 2)battery capacity:24V/8AH 3)battery type:Ni-HM 4)charging time:3-4h ...
[Keywords: electric bike, electric bicycle, city electric bicycle]
See All Items(13) from Cycleman Co.,Ltd [China]

Electric Bicycle Mould

We specialize in designing and making molds, such as plastic injection mold, die casting mold, blow mold, ...
[Keywords: e-scooter plastic parts die, electric bicycle plastic part, plastic parts mouldelectrice]
Taizhou Huangyan Weiyan Plastic Mould Factory [China]

Electric Bicycle(TDR237Z)

Technical parameters: 1)Packing size:165x32x98cm 2)Weight:70kg 3) Loading:75kg 4)Max speed:25km/h 5)Running ...
[Keywords: electric bicycles, electric motorcycle, electric scooter]
See All Items(2) from Shanghai Wangzhipai Vehicle Industry Co., Ltd [China]

Electric Bicycle T621

1) Wheel type/ size: Front + back aluminum 26" 2) Batteries: CE certified 36V-12Ah, SLA battery 3) Range ...
[Keywords: electric bicycle, electric bike, electric vehicle]
See All Items(6) from Shanghai Wind Rider Electric Bike Co.,Ltd [China]

Electric Bicycle
1)The frame is alloy 2)Six gear multi-speeds 3)26inch spoke wheel, M-finish rim 4) 36V, 10Ah lithium ...
[Keywords: Electric vehicle, Electric bicycle, bicycle]
See All Items(4) from Shandong Green.Tec Electric Technology Co.,Ltd [China]

Electric Bicycle YTLEB-9938
Motor: 180W brushless motor Battery: 36V12Aor 36v14A lead-acid battery Change time: 4-8 hours Max. speed: ...
[Keywords: electric scooter, electric bike, ATV]
See All Items(8) from Zhejiang Linan Industry and Trade Co.,Ltd [China]

Grand Electric Bicycle
Model TDR08Z-15 Maximum speed: ...
[Keywords: electric bicycle, battery]
See All Items(2) from Jiangsu New Continental Vehicle Co.,Ltd [China]

Electric Bicycle(BZ-1004)
BZ-1004 Powe: 250W Max load: 100kg Max speed: 25km/h Distance per charge: 45km Battery:36V/12Ah Input ...
[Keywords: electric bicycle, electric scooter, autobike]
Shanghai Benzhi Electric Bicycle Co.,Ltd.Shenzhen Branch [China]

Electric Bicycle
Zhejiang Crowd Power Co., Ltd. is located in Yongkang City, which is known as "the Science and Hardware ...
[Keywords: Electric Bicycle, Electric, Bicycle]
See All Items(8) from ZheJiang YongKang Crowd Power Co.,Ltd. [China]

Electric Bicycle Conversion Kits (DIY Kits)
Features For the competent mechanic our kit is available for self fit to your bicycle ! Easy fitment ...
[Keywords: Conversion Kit, BLDC Motor, electric bicycle]
See All Items(12) from Samhyun Co., Ltd. [Korea]

Electric Bicycle TDF36Z
Components Frame: Al Alloy 6061 T6 Tyres: 26"?.75",puncture resistant K-shield Rim: Al Alloy twin wall ...
[Keywords: alloy electric bicycke, electric bicycle, electric scooter]
See All Items(2) from Yongkang Chiyu Industrial Co., Ltd. [China]

Electric Bikes,E Bicycles
Electric City Bicycle Driving Mode: Electric Pedal with Assistance Weight: 28KG (battery included) Frame: ...
[Keywords: Electric Bikes, electric bicycles, city bike]
Lorstar International Corp. Ltd [China]

Electric Chopper Bicycle
48/350W; FRONT WHEEL: 26"; REAR WHEEL: 24" ...
Nanjing Skyland Co., Ltd [China]

Offline
Inquire now

Electric Bicycle(BZ-1004)
BZ-1004 Powe: 250W Max load: 200kg Max speed: 35km/h Distance per charge: 45km Battery:36V/12Ah Input ...
Shanghai Benzhi Electric Bicycle Co.,Ltd.Shenzhen Branch [China]

Offline
Inquire now

48V 1000W Electric Bicycle Conversion/Retrofit Kits
E-Bike conversion kits It's cool, it's fun, it's a joy of riding... It's still your bike Never being ...
See All Items(2) from LongFaith Group Co.,Ltd. [Hong Kong]

Offline
Inquire now

Wudi Electric Bicycle
Relative technical parameters of electric bike Model King of Loading (60V) Motor Model brushless Rated ...
See All Items(15) from Wudi Electrical Bicycle Corporation [China]

Offline
Inquire now

Electric Bicycle
Model No. TDR07666 Electric MOTOR 48V 350W,BRUSHLESS Battery 48V,12AH,4-CELLS,REMOVABLE Input Voltage ...
See All Items(14) from Jiangsu Autosun Vehicle Manufacturing Co.,Ltd [China]

Offline
Inquire now

Electric Bicycle
parameters list of QLM-13 vehicle with frame battery material:45# warranty:2 years wheel s dir specification:16 ...
See All Items(8) from Wuxi CELIMO Vehicle Manufacturing Co.Ltd [China]

Offline
Inquire now

Electric Bike,Electric Bicycle,E Bike,Bicycle
Frame Aluminum alloy 6061 T6 Tyres 26 ?? x 1.95, puncture resistant K-shield Rim Aluminum alloy twin ...
China Weiyi Industry & Trade.,Co Ltd [China]

Offline
Inquire now

Electric Bicycle
Power of motor :180/250W high speed brushless Accelerator: intelligent torque sensor 6-class Shimano ...
See All Items(5) from Zhejiang Huaheng Import and Export Co., Ltd [China]

Offline
Inquire now

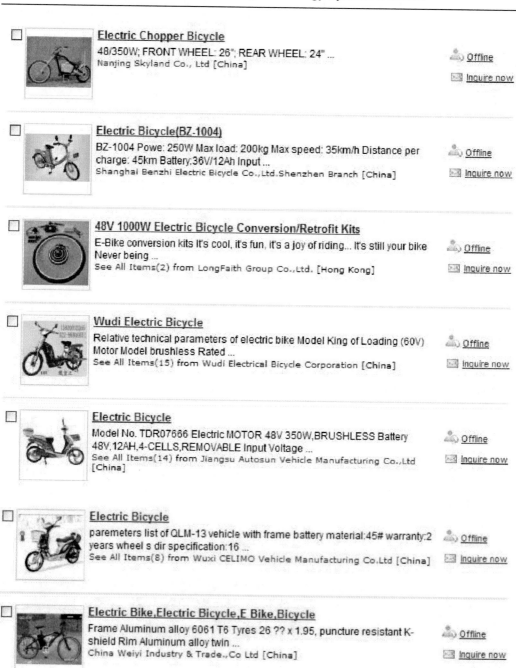

7 PCS Power LED/7 LED Bicycle Lighting
. 7 pcs power LED bicycle lighting: ABL-8007 . 7 LED bicycle lighting: ABL-8007A Primary Competitive ...
Zhejiang Angler Enterprise Co., Ltd [China]

Offline
Inquire now

E Bike Electric Bicycle
AT-D17 Specifications:26*1.75 Frame: Steel Frame Front fork: front shock absorberfork Brake: F /R ? ?rake Deraille ...
China RongKai Group-BST Bicycle [China]

Offline
Inquire now

Electric Bicycle(Li -bttery)
TYPE:TDR550Z SIZE:1080*480*920 WHEEL: 12" BRAKE: RR/DISC BATTERY:18V, 5.4Ah MOTOR:120W motor RECHARGE ...
See All Items(2) from Jiangsu Hongdou Imp&Exp Co.Ltd [China]

Offline
Inquire now

Hongjin Electric Bicycle Model: Tdr16z
"Hong Jin Electric Bicycle TDR16Z Technology Parameter" 1 Size 1730*450*1080 2 Weight 76kg 3 Tire 16*3.0 ...
See All Items(2) from Shanghai Hongjin Electric Bicycle Co., Ltd. [China]

Offline
Inquire now

Electric Bicycles with Alloy Frames Lithium Battery RPZ2002
Frame :6061ALLOY Lithium Battery 24V10AH Charging time 4-6h Motor 250w Brushless Geared Shimano 6Sp Net ...
See All Items(4) from Jinhua Repu Electric Scooter Co.,Ltd [China]

Offline
Inquire now

Electric Bicycles
Electric Bike - the benefits; No gasoline, no air pollution No license, no insurance No registration Easy ...
TBD Internationed Trade Compang [China]

Offline
Inquire now

Electric Bicycle
Control mode:pedal assistance/electric power Electric motor:brushless motor Rated power of motor:24V ...
See All Items(2) from Shanghai Rnager International Co.,Ltd. [China]

Offline
Inquire now

FZW-2 Electric Bicycle
FZW-2 CARTON SIZE 145X70X33 1. BODY SIZE (L X W X H) 1460 X 565 X 1,070MM 2. WHEEL BASE: 1075MM 3. GROUND ...
See All Items(6) from Wuxi Wanyu Electric Vehicle Co.,Ltd [China]

Offline
Inquire now

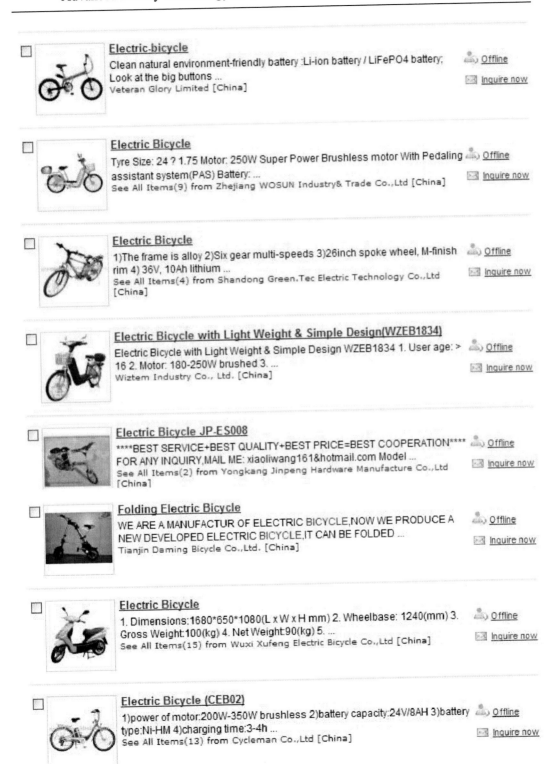

Electric Bicycle T621
1) Wheel type/ size: Front + back aluminum 26" 2) Batteries: CE certified 36V-12Ah, SLA battery 3) Range ...
See All Items(6) from Shanghai Wind Rider Electric Bike Co.,Ltd [China]

Offline
Inquire now

Grand Electric Bicycle
Model TDR08Z-15 Maximum speed: ...
See All Items(2) from Jiangsu New Continental Vehicle Co.,Ltd [China]

Offline
Inquire now

Electric Bicycle (CONCERT)
Features Model : CONCERT Model No. : SHEB11L Motor : 300W / BLDC Hub Geared Motor Running Mode : Auto ...
See All Items(12) from Samhyun Co., Ltd. [Korea]

Offline
Inquire now

Electric Bicycle
Zhejiang Crowd Power Co., Ltd. is located in Yongkang City, which is known as "the Science and Hardware ...
See All Items(8) from ZheJiang YongKang Crowd Power Co.,Ltd. [China]

Offline
Inquire now

Electric Bicycle YTLEB-9938
Motor: 180W brushless motor Battery: 36V12Aor 36v14A lead-acid battery Change time: 4-8 hours Max. speed: ...
See All Items(8) from Zhejiang Linan Industry and Trade Co.,Ltd [China]

Offline
Inquire now

CE UL Approved PAS Electric Bicycle Bike RLB-175A
RLB-175A Electric Bicycle electric bike 1)Frame Aluminium 2)Brush Motor Max Power 250W 3)Ni-MH Battery ...
See All Items(3) from Zhejiang Renli Vehicle Co.,Ltd [China]

Offline
Inquire now

Electric Bicycle
CE certification approvaled, motor power:200W,36V12Ah,Alloy frame, wheel base:26" ...
See All Items(3) from HODO Group Chituma Motorcycle CO.LTD [China]

Offline
Inquire now

Electric Bicycle
Products Name: Electric Bicycle Description: Motor: 180W Brushness Battery: 24V, 9 amps Li-ion Wheel ...
Zhejiang Aceme Electric Vehicles Co.Ltd [China]

Offline
Inquire now

Electric Bicycle
Dimension: 1450*300*85CM The Type of Motor: Brushless Motor The Output of Power: 350W Battery: 48V/10AH ...
Binyuan Group [China]

Offline
Inquire now

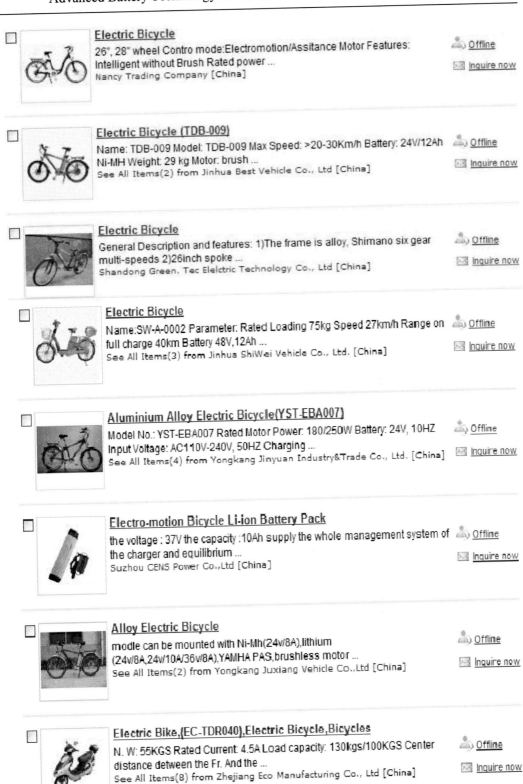

Electric Bicycle
TDR-44Z Wheel size: 16 ? * 3" Net Weight: 70.5kg Motor power: brushless Integrative brushless DC motor ...
See All Items(12) from Leisger Vehicle Manufactoring Co., Ltd [China]

Offline
Inquire now

Electric Bicycle
Motor Power: 500W with brush/brushless Dimensions(L*W*H): 1870*755*1030mm Speed: 40-50km/hr. Max load: ...
Wuxi YunLu Motorcycle Co.,Ltd [China]

Offline
Inquire now

Electric Bicycle
Size: 155*55*100(cm) N. W: 35KG G. W: 45KG Max speed: 28km/h Max range: 40km Max loading: 85kg Motor ...
See All Items(2) from HangZhou BenBao Electric Vehicle Industry Co.,Ltd [China]

Offline
Inquire now

Electric Bicycle
Detailed Product Description Specifications: 1) Wheel sizes: 18 x 2.125 2) Maximum load: 120kg 3) Top ...
See All Items(2) from Ningbo Pugongying Vehicle Technology Co.,Ltd [China]

Offline
Inquire now

Electric Bicycle
Exterior dimension: 1880 ? 55 ? 110mm Travel range on full charge:50~80km Motor Power:350W Brushless Wheel ...
See All Items(7) from ZheJiang Qun Ying Vehicle CO.,LTD [China]

Offline
Inquire now

Electric Bicycle
Electric Bicycle is our strong product. Our product is with below characters:convenient operation, fashion ...
Hebei Fulong Import & Export Co., Ltd. [China]

Offline
Inquire now

Electrical Bicycle
Product name:Steel folding electric bicycle 1)Motor power:250W brushless with gears 2)Controller:1:1(1:1 ...
Hangzhou Smile Import & Export Co., Ltd [China]

Offline
Inquire now

Electric Bicycle
our company has 12 economic entities, and all our products concluding Celimo electric bicycle, ompu ...
Wuxi Celimo Vehicle Manufacturing Co,.Ltd [China]

Offline
Inquire now

Electric Bicycles
Continued driving distance: 60/75 Vehicle weight: 105 kg Loading weight: 150 kg Max speed: 70km/h Rated ...
See All Items(3) from Zhejiang Haoren Electromechanical Co.Ltd [China]

Offline
Inquire now

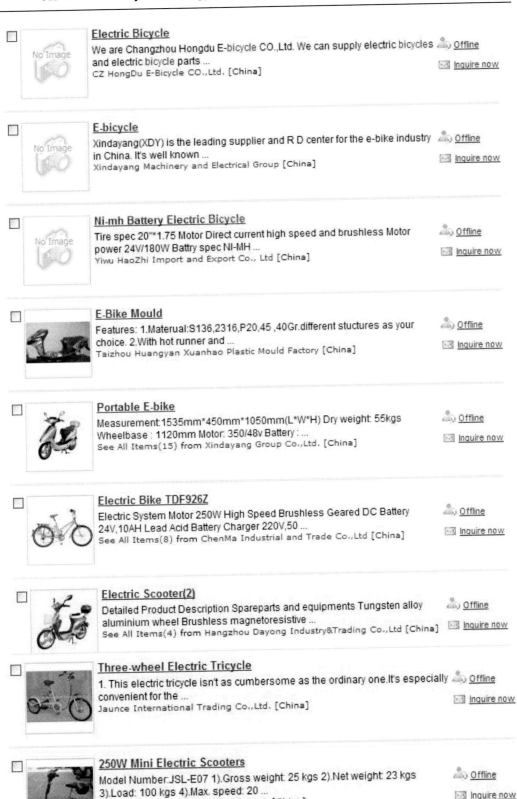

Electric Bike SE-TDL-20F01A-1
electric bike,Frame:alloy, Fork:suspension, Derailleur:SHIMANO 6 SPEED, Rim: 20"x1.5x12Gx36H ALLOY, ...
See All Items(9) from San Eagle Bicycle Manufactory Co., Ltd. [China]

Offline
Inquire now

Electric Bike Conversion Kit
Electric Bike Conversion Kit Type: Front wheel kit. Autonomy: 24 to 50 km kits with brushless motor. ...
Xingtai Evergreen Co.,Ltd [China]

Offline
Inquire now

Electric Bikes
Top Quality Electric Bicycles in China, supplied by ET Machinery Co., Ltd.We can supply all kinds of ...
ET Machinery Co., Ltd. [China]

Offline
Inquire now

Sight-seeing Carts
The Sight-seeing Electric Cars we supplied are designed in the internationally advanced technologies, ...
See All Items(8) from Shandong Senxin Trade Co., Ltd [China]

Offline
Inquire now

Electric Bike
Product size 186x60x113cm(LxWxH) N.W(with battery) 32.5kg Weight of Battery 4.75kg G.W 42kg Type of motor high ...
See All Items(2) from Shanghai New Success Imp&Exp. XCo.,Ltd [China]

Offline
Inquire now

Asia Star
ID: FB-JLQ-01 frame aluminous alloy motor type brushless with tooth high speed motor power: 180 w dimension ...
Hangzhou Jialiqi Bicycle Manufacture Co.,Ltd. [China]

Offline
Inquire now

Tdr130z
Dimension(L*W*H) 1550 600 1030MM Max. speed 28 1 km/h Max.load Capacity: 75kg Min charge range 45km Battery: 48V/ 1 0HA 4pc/set Charge 48V 1.8A Wheel ...
Daktra Co.,Ltd [China]

Offline
Inquire now

Electric Bike
Brand: PRIEST Model NY-013 Frame material: aluminum alloy Wheels: Alloy rims and lightweight alloy hubs Brake: ...
See All Items(3) from CiXi NouYa Electric Appliance Co., Ltd [China]

Offline
Inquire now

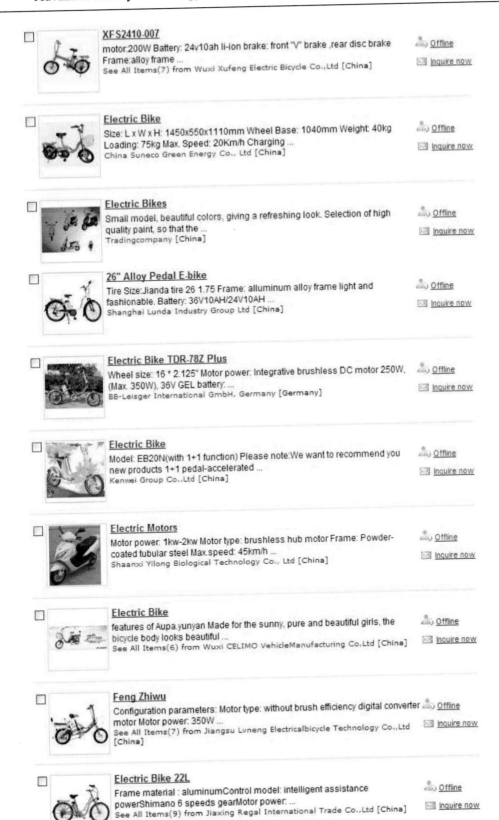

E Bike Motor

Suzhou Bafang Motor Science Co., Ltd is high science and technology enterprise and professional motor ...

Suzhou Bafang Electric Motor Science-technology Co.,Ltd [China]

Offline
Inquire now

E-bike Folding with Li Battery

Very New arrival ,Mini design it's very handiness,and easy to folding release,Equiped pocket Lithium ...

ForU Houseware Co.,LTD [China]

Offline
Inquire now

Persesus

Specification:18",48V,350W Range:25-40km ...

See All Items(4) from Shenzhen China Bicycle Company (Holdings) Limited [China]

Offline
Inquire now

CE Approved Electric Bike

? Power of motor:220W brushless ? Battery:36V,10AH LI-ION ? Charging time:5-6h ? Charger life: >500 times ? Climbing ...

See All Items(2) from China XinXin Stainless Steel Casting Inc. [China]

Offline
Inquire now

Electric Bike

Power of motor:180-250W brushless ? Battery:10AH/24V,or 10AH/36V Li-ion ? Charging time:5-6h ? Charger ...

Zhejiang Harvest Industrial CO.,LTD. [China]

Offline
Inquire now

Zaizhongwang

L? ?:1800 620 1140mm Battery: 10V/10AH ? PCS Input voltage(V):110--- 220V Output voltage(V):60V Charge ...

See All Items(4) from Wu Xi Lucky Lion Vehicle Industry Co.,Ltd [China]

Offline
Inquire now

Only 20lbsThe Most Lightweight and Convenient Electric Bike

Features: 1.Unique & compact design.Only 19.9lbs,The Most Lightweight And Convenient Electric Scooter ...

Shijiazhuang Further Imp.&Exp. Ltd.,Co. [China]

Offline
Inquire now

Silicone Power Battery

12v10ah(2HR) 4.2kg, 151*98*95*100 Features of the silicone power battery: * Fast charging ability * High ...

Greensaver Corporation [China]

Offline
Inquire now

Electric Scooter(Xgy)

Parameter of e-scooter 1. Dimensions:1440*650*1090(L x W x H mm) 2. Wheelbase: 1,250(mm) 3. Gross Weight:95(kg) 4. ...

See All Items(2) from Wuxi Xufeng Electric Bicycle Co.,LTD [China]

Offline
Inquire now

Lady Electric Bike
PRINCESS Function: power assistant Frame 26 alloy Fork: Suspention alloy Brake: front and rear V-brake Battery :36v ...
Jiangmen Huanan Special Vehicle Industrial Co Ltd [China]

    Offline

    Inquire now

Model No : TDN802Z / TDN802ZS
/ Remark: The Model No added S is with SHIMANO SPROCKET 6 GEAR and front/rear disc brake(6) frame ...
Zhejiang Xinghai Energy Technology Co.,Ltd [China]

    Offline

    Inquire now

City Tango
CITY TANGO Frame: Aluminum alloy Size:26?? x 1.95, puncture resistant K-shield Motor:BLDC GEARED Motor,36v ...
Tianjin Kaite Bike Co.,Ltd [China]

    Offline

    Inquire now

Electrical Bike
we are a professional manufacturer of electric bicycles and scooters. It locates in Ningbo, China, about ...
Ningbo Wozom Industry&Trade Co.Ltd [China]

    Offline

    Inquire now

APPENDIX D. INTERVIEW QUESTIONS AND DISCUSSION TOPICS

The following questions were submitted to organizations that develop lithium-ion batteries in the People's Republic of China two weeks before the author was to meet with them.

1. What is the status of your battery technology? What is the level of development — cell, module, or full pack?
2. What are your battery technology applications? What are their power, energy, volume, and weight?
3. What testing methodology do you use? What are the charge/discharge limitations, cycle life, capacity, voltage, temperature operating range, and effects of aging?
4. How many units do you produce per year? Where are your manufacturing facilities, and what equipment do you use?
5. Are any special feature(s) available on your current products?
6. Are raw materials available domestically as well as imported?
7. Where are your batteries sold? Are they sold domestically or exported? Are you working with any other companies overseas or domestically? How many batteries are produced for domestic use and for export purpose?
8. What is the cost of your batteries domestically and overseas?
9. Is your battery cell, module, or pack available for testing and evaluation if we can work out a confidentiality agreement? Could we evaluate your technology at

Argonne National Laboratory in the United States? When can you make the battery available for evaluation?

10. Are you conducting research and development (R&D) at your facilities or with companies in China or overseas to improve your products? Are you developing new products or technology? In general, what is the nature of your agreements? Is your company participating in any joint ventures or equity partnerships?
11. Currently, are you working on batteries for electric vehicles (EVs), hybrid electric vehicles (HEVs), and/or plug-in hybrid electric vehicles (PHEVs)?
12. Are you interested in developing batteries for EVs, HEVs, and PHEVs?
13. What is the size of your company? How many employees do you have? What are your sales per year in terms of kWh of capacity sold or in value sold in yuan?
14. Do you sell batteries directly as retail products? Do you sell batteries to other companies that convert them into packs with controllers? Is quality control causing you problems in selling batteries to some potential customers?
15. What are the incentives are offered by the Chinese Government to battery developers? Are these same incentives available to domestic and/or international companies working with Chinese companies?
16. Do you have intellectual property, such as patents, joint venture agreements, or other rights, to protect your products? How important are they with regard to developing new products versus improving products and with regard to new ventures versus current manufacturers? Do you purchase battery technologies and specialize only in production, or do you invest in battery R&D to develop your own products? If you do your own R&D, how much do you spend per year?

APPENDIX E. COMPANIES MARKETING ELECTRIC BICYCLES WORLDWIDE

Alien Scooters

- **Business type:** Retail sales
- **Product types:** Electric bicycles, electric scooters, mopeds, custom electric bicycles and electric bicycle components and accessories, solar electric charging systems, electric bike conversion kits, electric motorcycles, training for light electric vehicle technicians, and electric bike tours and rentals
- **Service types:** Bicycles and all electric vehicles
- **Address:** 1122 S. Lamar Blvd., Suite B, Austin, Texas 78704
- **Telephone:** 512-447-4220
- **Fax:** 512-444-8687

Alternative Vehicle Distributors, Inc.

- **Business type:** Manufacturer and distributor
- **Product types:** Zem zero-emission machines, Maximo tandem buggies

- **Service types:** Sales and marketing
- **Address:** 1530 West 10th Place, Tempe, Arizona 85281
- **Telephone:** 480-505-0308
- **Fax:** 480-966-4422

Battery Bikes

- **Business type:** Retail sales
- **Product types:** Electric scooters
- **Address:** 2894 Superior Drive, Livermore, California 94550
- **Telephone:** 877-860-3900, ext. 532640 (toll free)
- **Fax:** 925-961-0967

Currie Technologies Inc.

- **Business type:** Designer, engineer, manufacturer, importer, and distributor
- **Product types:** Ezip and Izip hybrid electric bikes and scooters, mountain bikes and urban cruisers, electric propulsion systems, conversion kits for bicycles, parts and accessories, bikes with various platforms (twist and go or pedal assist [TAG/PAS], torque measurement method [TMM], and Evo-drive technology)
- **Address:** 9453 Owensmouth Ave, Chatsworth, California 91311
- **Telephone:** 818-734-8123
- **Fax:** 818-734-8199
- **Web site:** http://www.currietech.com

CityBug USA

- **Business type:** Manufacturer
- **Product types:** Electric scooters
- **Address:** 1060 Commerce Blvd. North, Sarasota, Florida 34243
- **Telephone:** 888-743-3738
- **Fax:** 941-351-2699

CityGlide

- **Business type:** Retail sales
- **Product types:** Electric vehicle conversion kits, electric bicycles, electric scooters
- **Address:** 4108 Norcross, Plano, Texas 75024
- **Telephone:** 214-335-8500

CityMoped.ca

- **Business type:** Retail sales, c
- **Product types:** Electric scooters, electric motorcycles, electric bicycles
- **Service types:** Consulting, installation, maintenance, and repair
- **Address:** 4060 Salal Drive, Nanaimo, British Columbia, Canada V9T 5J7
- **Telephone:** 250-740-3953
- **Fax:** 250-740-3953

Consortium Manufacturing, Inc.

- **Business type:** Manufacturer
- **Product types:** Electric bicycle components, hybrid electric vehicles, fuel cell system components, electric scooters, meters and measuring equipment, geothermal energy system components, precision sheet metal and machined parts, fabricators for high-tech industries
- **Address:** 5730 West 108th Place, Westminster, Colorado 80020
- **Telephone:** 720-244-5453
- **Fax:** 303-466-7165

Crave Sports, Inc.

- **Business type:** Manufacturer, wholesale supplier
- **Product types:** Electric bicycles, electric scooters, electric tricycles, adult tricycles, electric skateboards
- **Address:** 5511 Ekwill St., No. D, Santa Barbara, California 93111
- **Telephone:** 805-967-2216
- **Fax:** 805-967-2752

Delaware Direct Sales, Inc.

- **Business type:** Retail sales
- **Product types:** Electric scooters
- **Address:** 58 Clinton Street, Delaware City, Delaware 19706
- **Telephone:** 302-838-0656

Doran Motor Company

- **Business type:** Manufacturer, distributor
- **Product types:** Lightweight electric vehicles, electric all-terrain vehicles, electric tractors, three-wheeled electric vehicles (no sales, parts, or repairs for two-wheeled scooters)
- **Service type:** Engineering
- **Address:** 5842 McFadden Avenue, Unit R, Huntington Beach, California 92649
- **Telephone:** 714-377-7776

Drive-Electric Solutions – Village Energy

- **Business type:** Retail electric vehicle sales, consultants, publisher, Internet service provider (ISP)
- **Product types:** eGO cycles, Wild Ride and Stealth electric bike conversion kits, Village Energy custom t-shirts
- **Service types:** Drive-Electric.com ISP, publisher of *New Energy News,* consultants on alternative energy and electric vehicles
- **Address:** 300-108A Carlsbad Village Drive No. 237, Carlsbad, California 92008
- **Telephone:** 760-729-8075 or 760-580-0075
- **Fax:** 240-337-8567

Ecomotion

- **Business type:** Retail sales
- **Product types:** Alternative fuel vehicles, electric motorcycles, 100% electric scooters, 100% electric cars, 100% electric trucks, 100% electric all-terrain vehicles, diesel vehicles, used diesel automobiles, hybrid vehicles, pre-owned U.S. Environmental Protection Agency (EPA) Smartway-certified vehicles, Zap, Toyota, Honda, Miles, Myers, Dymac, Bravo
- **Address:** 1625 NE Sandy Blvd., Portland, Oregon 97232
- **Telephone:** 503-244-5658
- **Web site:** http://eco-motion.com/
- **E-mail:** Send to Ecomotion: Earth Friendly Vehicles

Enviro-Bike Electric Vehicles

- **Business type:** Retail sales
- **Product types:** Electric scooters, electric wheelchairs, electric all-terrain vehicles, electric dirt bikes, electric motorcycles, electric scooters, electric chariots, electric bicycles, electric tricycles, electric vehicle components, electric vehicle conversion

kits, mobility products, and parts and accessories for all of its electric vehicles, Palmer electric vehicles (Twosome and Joyrider), Schwinn electric Stingray
- **Service types:** Maintenance and repair services
- **Address:** 347 Encinitas Blvd., Suite A, Encinitas, California 92024
- **Telephone:** 760-722-1146
- **Web site:** http://www.enviro-bike.com/

E-Cycle Electric Vehicles

- **Business type:** Manufacturer, retail sales, wholesale supplier, exporter, and importer
- **Product types:** Electric scooters, electric bicycles, electric motorcycles, electric bicycle components, electric vehicles, electric cars, folding bicycles, exclusive distributor of EVT 168 and EVT 4000e, Canada's only TransportCanada-approved limited-speed motorcycles
- **Service types:** Research and development of electric drive systems
- **Address:** 1703 W. 4th Avenue, Vancouver, British Columbia, Canada V6J-1M2
- **Telephone:** 1-866-309-6717
- **Web site:** http://www.e-cycle.ca/

eZee SA

- **Business type:** Wholesale supplier, importer
- **Product types:** Advanced electric bicycles, electric scooters
- **Address:** Johannesburg, ZA, South Africa
- **Telephone:** +27 827454962
- **Web site:** http://www.ezeebike.co.za/

e-BikeKit.com

- **Business type:** Retail sales
- **Product types:** Electric bicycle kits, electric scooters
- **Address:** 11091 SW Springwood Drive, Tigard, Texas 97223
- **Telephone:** 503-887-7783

E-go Personal Transport

- **Business type:** Retail sales
- **Product types:** Electric bicycles, electric scooters, electric bicycle components
- **Address:** 52 High Street, Marlborough, Wiltshire, United Kingdom SN8 1HQ
- **Telephone:** 07974 723996

E-Ride

- **Business type:** Engineering, wholesale supplier, importer, retail sales
- **Product types:** Motorino electric motorcycles, Motorino electric scooters, electric bicycles, electric skateboards, special-purpose electric vehicles, electric conversions
- **Address:** 240 East 2nd Avenue, Vancouver, British Columbia, Canada V5T 1B7
- **Telephone:** 1-604-331-0555

E.S. Buys, Inc.

- **Business type:** E-commerce
- **Product types:** Gift items, gadgets, holiday and birthday gifts, home and living items (pillows, lamps, etc.), healthy living products (bath scales, exercise equipment, beauty aides, etc.), home security items, spy technology items (listening devices, video cameras), juke boxes, music boxes, Crosley radio products, small nostalgic appliances (popcorn poppers, snow cone machines, etc.), sports and hobby items, Stern pinball machines, full-size arcade games, toys and recreational items (foosball tables, air hockey tables, radio-control toys, folding mountain bikes), travel and automotive items, watches and clocks
- **Address:** P.O. Box 783, Newbury Park, California 91320
- **Telephone:** 805-499-1141

EarthFriendlyMachines.com

- **Business type:** Retail sales
- **Product types:** Electric bicycles, electric scooters, electric motorcycles
- **Address:** 5440 SW Westgate Drive, Suite 320, Portland, Oregon 97221
- **Telephone:** 800-286-8875

Eco-lectric

- **Business type:** Manufacturer, retail sales, wholesale supplier, exporter
- **Product types:** Electric scooters, electric motorcycles
- **Address:** Systems House, Rotherside Road, Sheffield, Derbyshire, United Kingdom S21 4HL
- **Telephone:** 01246 431431

Edge-EV

- **Business type:** Manufacturer, retail sales, wholesale supplier
- **Product types:** Electric bicycles, sealed lead acid batteries, DC to AC power inverters, electric scooters
- **Address:** 81 Francis Street, Cambridge, Ontario, Canada N1T-1B4
- **Telephone:** 519-621-7338

Ekovehicles Pvt. Ltd.

- **Business type:** Manufacturer, retail sales, importer
- **Product types:** Electric scooters, electric vehicles
- **Service types:** Engineering, research, maintenance, and repair
- **Address:** No.10, Brunton Road, M.G. Road Cross, Bangalore, Karnataka, India 560 025
- **Telephone:** 091-080-41240814/815
- **Fax:** 091-080-41 142013

EleBike Center Sweden

- **Business type:** Retail sales, wholesale supplier, exporter, importer
- **Product types:** Pogo sticks, electric kick scooters, minibikes, electric bicycles, electric scooters, SurfScooters, AB slides, sports equipment
- **Address:** Kapellgatan 24, S-265 36, Sweden
- **Telephone:** +46-708-855439 +46-42-55439
- **Fax:** +46-708-85543 8

Electric Bikes Northwest

- **Business type:** Retail sales
- **Product types:** Electric bicycles
- **Service types:** Full service department
- **Address:** 110 North 3 6th, Seattle, Washington 98103
- **Telephone:** 206-547-4621

Electric Clean Air Cycles

- **Business type:** Wholesale supplier, exporter
- **Product types:** Electric scooters, electric motorcycles, electric bicycles, electric vehicle components
- **Address:** 3765 Delta Circle, Corona, California 92881

- **Telephone:** 909-818-0310
- **Fax:** 603-804-8296

Electric Coast Vehicles Inc.

- **Business type:** Manufacturer, retail sales, importer
- **Product types:** Electric bicycles, electric motorcycles, electric vehicle conversion kits, electric scooters
- **Service types:** Consulting, design, engineering, project development, research, maintenance, and repair
- **Address:** 238 East Esplanade, North Vancouver, British Columbia, Canada V7L 1A3
- **Telephone:** 604-985-1615

Electric Cyclery

- **Business type:** Retail sales
- **Product types:** Electric scooters, electric bicycles, electric vehicle batteries, alternative fuel vehicles, electric motorcycles, electric vehicle components, Bionx motor kits, Optibikes, Goped scooters, Jackal electric bikes, eGO cycles
- **Address:** 900 N. Coast Highway, Laguna Beach, California 92651
- **Telephone:** 949-715-2345

Electric Motorsport

- **Business type:** Manufacturer, retail sales, wholesale supplier, exporter, importer
- **Product types:** Electric motorcycles, electric scooters, electric vehicle batteries, electric vehicle components, hub motors
- **Address:** 2400 Mandela Parkway, Oakland, California 94607
- **Telephone:** 510-839-9376
- **Fax:** 510-832-7010

Electric Ride

- **Business type:** Retail sales
- **Product types:** Electric scooters
- **Address:** P.O. Box 460543, Saint Louis, Missouri 63146
- **Telephone:** 314-563-0304
- **Fax:** 314-298-9699

Electric Rides

- **Business type:** Retail sales
- **Product types:** Alternative fuel vehicles, electric scooters, electric vehicle components, electric vehicles, electric bicycles, rider accessories
- **Address:** 415 NE Santiam Blvd. (mail: 324 NE 4th Ave.), Mill City, Oregon 97360
- **Telephone:** 503-897-3152

Electric Sierra Cycles

- **Business type:** Manufacturer, wholesale, retail sales
- **Product types:** Synergy Cycle electric bike
- **Address:** 302 Pacific Ave., Santa Cruz, California 95060
- **Telephone:** 1-831-425-1593 (toll free 1-877-372-8773)
- **Fax:** 1-831-425-5988

Electric Transportation Solutions, LLC

- **Business type:** Retail sales, wholesale supplier, exporter, importer
- **Product types:** Electric scooters, electric bicycles, electric cars, electric golf carts, solar electric power systems, fuel cell systems
- **Address:** 100 South Bedford Road, Suite 340, Mt. Kisco, New York 10549
- **Telephone:** 866-895-2238

Electric Vehicle Technologies Inc.

- **Business type:** Manufacturer
- **Product types:** Electric bicycles, electric motorcycles, electric scooters, alternative fuel vehicles
- **Address:** 7320 North Linder Ave., Skokie, Illinois 60077
- **Telephone:** 847-673-2718
- **Fax:** 847-675-1827

Electric Vehicles (Thailand) Co., Ltd.

- **Business type:** Manufacturer, retail sales, wholesale supplier, exporter
- **Product types:** Electric bicycles, electric scooters, electric cars, electric vehicle components
- **Service types:** Maintenance and repair
- **Address:** 374, BNK Building, Rama 4 Road, Mahapreuttaram, Bangrak, Bangkok, Thailand 10500

- **Telephone:** (66-2) 236-2020, ext. 302
- **Fax:** (66-2) 237-2002

Electric Vehicles Ireland

- **Business type:** Retail sales, wholesale supplier, importer
- **Product types:** Electric bicycles, electric scooters, electric vehicle conversion kits, electric cars
- **Address:** 33 Strandville Gardens, O'Callaghan Strand, Limerick, Ireland
- **Telephone:** +353- 087-786963 1

Electric Wheels Inc.

- **Business type:** Retail sales
- **Product types:** Electric cars, electric scooters, electric bicycles, golf cars, golf carts, neighborhood electric vehicles, low-speed vehicles, mobility vehicles and chairs, lift kits
- **Address:** 1555 12th St. SE, Suite 110, Salem, Oregon 97302
- **Telephone:** 503-485-0588
- **Fax:** 503-485-0590

Electrik Motion

- **Business type:** Retail sales, wholesale supplier
- **Product types:** Electric bicycles, electric scooters, electric vehicle conversion kits, electric bicycle components, electric vehicle components
- **Address:** 7 Tamarac Ave., New York, New York 10956
- **Telephone:** 866-372-6687

Electro Ride Bikes and Scooters

- **Business type:** Retail sales
- **Product types:** Electric bicycles, electric scooters, electric bicycle components
- **Address:** 2807 Jones Ave., Milpitas, California 95035
- **Telephone:** 408-262-8975

Elite Industry

- **Business type:** Wholesale supplier, importer
- **Product types:** All-terrain vehicles, motor scooters, dirt bikes, motorcycles, electric scooters, mobility scooters, portable generators
- **Address:** 150 Commerce Way, Walnut, California 91789
- **Telephone:** 909-595-0850

Eloiz Tomola S.L.

- **Business type:** Retail sales, wholesale supplier, exporter, importer
- **Product types:** Electric scooters, electric bicycles
- **Address:** Hacienda Las Chapas, Avda 9, 175 Marbella, Malaga, Spain 29600
- **Telephone:** +34 952833564

Emerging Vehicles

- **Business type:** Retail sales
- **Product types:** Electric bicycles, electric scooters
- **Address:** One Broadway No. 1400, Kendall Square, Cambridge, Massachusetts 02142
- **Telephone:** 617-583-1393
- **Fax:** 617-758-4101

en-Trade International Corp. Shanghai

- **Business type:** Manufacturer, wholesale supplier, exporter
- **Product types:** Electric bicycles, electric motorcycles, electric scooters, garden items that run on rechargeable batteries
- **Address:** Room 101, No. 58, Lane 168, Qin Chen Road, Minhang Area, Shanghai, China 200011
- **Telephone:** +86-21-54959017
- **Fax:** +86-21-54959019

EPS, Energy Propulsion System Inc.

- **Business type:** Manufacturer, retail sales, wholesale supplier, exporter
- **Product types:** Electric bicycles, electric bicycle components, electric scooters, small wind turbines
- **Address:** 73, St. Georges North, Asbestos, Quebec, Canada J1T 3M7
- **Telephone:** 819-879-0041 ERDM Solar

- **Business type:** Manufacturer, provider of complete solutions, wholesale supplier, exporter, importer Product types: Produce ERDM solar photovoltaic modules (made with Q cells from Germany); supply solar charge controllers (Outback, Steca), deep-cycle batteries from Surrette, DC/AC power inverters from Outback Power Systems, small wind turbines from Southwest Windpower, stand-alone power systems (island) for remote homes or ranches, grid-tied systems (connected to the Internet), hybrid systems (wind and solar), uninterruptible power supply (UPS) systems; provide turnkey solutions for homes, businesses, communities, and industries
- **Service types:** Installation
- **Address:** Jinete 21 Fraccionamiento El Rodeo, San Andres Tuxtla, Veracruz, Mexico 95765
- **Telephone:** 52-294-9427520
- **Fax:** 52-294-9427524

Esarati Electric Technologies Corp.

- **Business type:** Manufacturer
- **Product types:** Electric motorcycles, electric scooters, electric vehicles
- **Address:** 10900 NE 8th St., Suite 900, Bellevue, Washington 98004
- **Telephone:** 877-843-1989
- **Fax:** 425-990-5981

Esco Sport Product Co., Ltd.

- **Business type:** Manufacturer
- **Product types:** Electric scooters, electric bicycles
- **Address:** 88 Heng Feng Road, Yongkang, Zhejiang, China 315300
- **Telephone:** 0086-571-85584969

EscooterPro

- **Business type:** Manufacturer, exporter
- **Product types:** Electric scooters, electric vehicle batteries, electric bicycles, electric vehicle components
- **Address:** Room 1113, 1 1/F, Wellborne Community Centre, 8 Java Road, Hong Kong, China
- **Telephone:** (852) 25120093
- **Fax:** (852) 28073287

EV Tech

- **Business type:** Retail sales
- **Product types:** Electric bicycles, solar electric power systems, alternative fuel vehicles, electric cars
- Address: 4310 Wiley Post Rd. No. 217, Addison, Texas 75001
- Telephone: 972-851-9990
- Fax: 972-851-9993

EVdeals

- **Business type:** Manufacturer, retail sales
- **Product types:** Electric scooters, electric bicycles, electric bicycle components, electric vehicle components, electric vehicle conversion kits, sealed lead acid batteries, Currie Technologies US Pro Drive (USPD) conversion kits, brushless motors, Brushless Motor Corporation (BMC) brushless motors, MAC Motors Ltd. motors, gears, custom sprockets, and belt drives
- **Service types:** Design, installation, maintenance, repair, testing
- **Address:** 1032 North St., Plainville, Massachusetts 02762
- **Telephone:** 508-695-3717
- **Fax:** 508-643-0233

EVT Technology Co., Ltd.

- **Business type:** Manufacturer, exporter
- **Product types:** Electric scooters
- **Address:** 66 Hwa-Ya 1 Road, Hwa-Ya Technical Park, Kuei-Shan Hsiang, Taoyuan Hsien Taiwan, China 333
- **Telephone:** +886-3-397-0022
- **Fax:** +886-3-397-2200

EVtransPortal.com

- **Business type:** Market research and consulting in advanced transportation technology
- **Product types:** Electric cars, batteries, electric vehicles, lithium polymer batteries, electric vehicle components, electric motorcycles, electric scooters, electric vehicle drive systems, electric vehicle propulsion systems, electric vehicle parts, hybrid buses, hydrogen fuel cell vehicles, lithium ion batteries
- **Service types:** Consulting, project development, research, site surveys, assessments
- **Address:** 6444 N. Glenwood Ave., Chicago, Illinois 60626

Express Power Products

- **Business type:** Wholesale supplier, importer
- **Product types:** Electric scooters, electric mobility scooters
- **Address:** 9615-111 Ave., Edmonton, Alberta, Canada T5G 0A9
- **Telephone:** 780-453-3754

Green Scene

- **Business type:** Retail sales
- **Product types:** Electric cars, electric bicycles, electric scooters, electric motorcycles, hybrid electric vehicles
- **Address:** 14221 SE McLoughlin, Milwaukie, Oregon 97267
- **Telephone:** 503-659-6622
- **Web site:** http://www.thegreensceneev.com

Green Machines LLC

- **Business type:** Retail sales, wholesale supplier
- **Product types:** Electric scooters, electric bicycles
- **Address:** 5333 SW 75th Street, V131, Gainesville, Florida 32608
- **Telephone:** 352-871-6725

Hills Motors

- **Business type:** Retail sales
- **Product types:** Electric bicycles, electric scooters
- **Address:** 4011 Pacific Blvd., San Mateo, California 94403
- **Telephone:** 650-573-7425
- **Fax:** 650-573-8721

Numotion Inc.

- **Business type:** Wholesale supplier, importer
- **Product types:** Electric scooters, E-vehicle components, electric bicycles, recreational vehicle power systems, electric vehicle conversion kits, sealed lead-acid batteries
- **Address:** 90 Halcyon Drive, Bristol, Connecticut 06010
- **Telephone:** 860-585-6122, ext. 106

Oceanline Scooters & Mobility

- **Business type:** Retail sales, wholesale supplier
- **Product types:** Electric scooters, electric bicycles
- **Address:** 9043 Crawfordsville Road, Indianapolis, Indiana 46234
- **Telephone:** 317-290-0450

Power Assist Products

- **Business type:** Retail sales
- **Product types:** Electric bicycles, electric scooters, electric cars
- **Address:** 1008 Warwick Drive, Macon, Georgia 31210
- **Telephone:** 912-971-4624
- **Fax:** 912-757-8549

R Martin Bikes

- **Business type:** Retail sales, wholesale supplier, importer
- **Product types:** Plug-in electric bicycles, electric motorcycles, electric mopeds, electric scooters, plug-in electric
- **Service types:** Research, maintenance, repair
- **Address:** 2125 Goodrich Ave., Suite B, Austin, Texas 78704
- **Telephone:** 877-680-8400
- **Web site:** http://www.rmartinbikes.com/

Rao Services Inc.

- **Business type:** Exporter, importer
- **Product types:** Alternative fuel vehicles, backup power systems, electric scooters, electric motorcycles, large hydro energy system components, remote home power systems, gas-electric generators, hydrogen-based electric and home heating generators, other fuel-saving products
- **Service types:** Consulting, project development, financial
- **Address:** 2937 41st Ave., 2nd Floor, Suite One, Long Island City, New York 11101
- **Telephone:** 718-726-0411
- **Fax:** 718-726-2118

Synergy Cycle

- **Business type:** Manufacturer, retail sales, wholesale supplier, importer, exporter
- **Product types:** Electric bicycles, the Synergy cycle, electric scooters, electric bicycle components, lithium batteries, lead acid sealed and gelled batteries, iron phosphate batteries, starting lead acid (SLA) batteries, motors for kit installation
- **Address:** 302 Pacific Ave, Santa Cruz, California 95060
- **Telephone:** 831-425-1593 sales, 877-372-8773 service
- **Fax:** 831-425-5988
- **Web site:** http://www.synergycycle.com

Santa Barbara Electric Bicycle Company

- **Business type:** Manufacturer, retail sales, wholesale supplier
- **Product types:** Electric bicycles, electric scooters, electric motorcycles, electric off-road vehicles
- **Address:** 630 Anacapa Street, Santa Barbara, California 93101
- **Telephone:** 866-675-7792
- **Fax:** 805-275-2338

USA-Bike

- **Business type:** Retail sales
- **Product types:** Electric bicycles, electric bicycle components, electric scooters, electric bicycle retrofit kits
- **Address:** P.O. Box 653, Willoughby, Ohio 44096
- **Telephone:** 440-975-9820
- **Fax:** 440-975-9820

X-Treme Electric Scooters

- **Business type:** Manufacturer, wholesale supplier, drop shipper for online businesses
- **Product types:** Electric scooters, electric bicycles, electric mopeds, electric hybrid bicycles, hybrid mopeds, all-terrain vehicles, sea scooters, gas scooters, pocket bikes, x-treme, scooter parts, mobility scooters
- **Service types:** Wholesale supplier, drop shipping
- **Address:** 910 N. 19th Ave. E., Newton, Iowa 50208
- **Telephone:** 402-603-4445
- **Fax:** 641-787-9221
- **Web Site:** http://www.x-tremescooters.com/

ZAP Manufacturers

- **Business type:** Manufacturer, distributor, web retail sales, wholesale supplier
- **Product types:** Electric cars, electric trucks, electric bicycles, electric tricycles, electric scooters, electric motorcycles, electric sea scooters, electric all-terrain vehicles, other low-power electric vehicles, hybrid cars, fuel cell cars, smart cars, advanced transportation cars, ZAP cars, alcohol-powered vehicles, ethanol cars, high-mpg (miles per gallon) cars, Xebra electric cars
- **Service types:** System design
- **Address:** 501 Fourth Street, Santa Rosa, California 95401
- **Telephone:** 707-525-8658
- **Fax:** 707-525-8692
- **Web site:** http://www.zapworld.com

APPENDIX F. SUMMARY OF THE REPORT
MARKET RESEARCH ON POWER LI-ION BATTERY IN CHINA 2007

Market Research on Power Li-ion Battery in China 2007 was prepared by the China Social Economic Investigation & Research Center, Beijing, China, and issued on March 29, 2007. The original report is in Chinese.

The first part of the report describes the basic knowledge of lithium-ion batteries and other power batteries, such as lead-acid, nickel-metal hydride, nickel cadmium, and fuel cells. The report points out that lithium-ion power batteries have zero emissions, zero pollution, high energy density, and high cycle life, and these characteristics reflect the expectations related to replacing lead-acid and nickel-metal hydride battery systems in the development and application of power batteries all over the world.

The market analysis of the lithium-ion power battery can be summarized as follows:

- In China, the market for a small capacity of lithiun-ion batteries for cell- phones and laptops is being saturated. The large capacity of lithium-ion power batteries has not entered the market yet.
- The lithium-ion power battery, however, is good enough to be applied in electric bikes and electric motorcycles. The electric motorcycle made by Taiwan EVT Electric Motorcycles Company employs a 36-V/1 00-A•h lithium-ion battery. Its driving range reaches 200 km and has a speed of up to 90–1 00 km/h.
- In the area of electric automobiles, solid polymer lithium-ion batteries may play a main role in the coming 2–3 years, as compared with the liquid lithium- ion battery. The energy density of the former is 30% higher than that of the latter. The shape of the former is more flexible, which could be strip-like, cylindrical, and prismatic.
- In the area of electric motorcycles, there are mainly three kinds of power batteries: lead-acid, nickel-metal hydride, and lithium-ion. So far, nickel-metal hydride is an ideal power source for electric motorcycles. Issues affecting lithium-ion batteries must be overcome (i.e., fast charging, large current, consistency between unit cells, and safety) before these batteries can be used in electric motorcycles. The

- requirement for protective electronic boards, a battery management system, and a thermal management system would increase the cost of the lithium-ion power battery.
- Production of electric bikes started in 1998 — at that time, there were 16 companies, and annual production was 58,000 in China. Since then, the industry of electric bikes has been developing very rapidly; annual growth was 87% on average. In 2006, 19.5 million electric bikes were sold. The competition between electric-bike companies intensifies daily. Survival in the market is difficult for companies with annual sales lower than 20,000– 30,000 bikes.
- The lithium-ion power battery pack for electric bikes consists of multi-unit cells in series or mixed-in series and parallel. The problem is that consistency for each unit cell is hard to maintain, which impacts the safety and life of the whole battery pack. Thus, it is critical to keep the lithium-ion battery 100% safe, if a good system for managing the lithium-ion battery pack could be invented.
- Over 300 companies make over 100 million power tools of over 200 kinds. In coming years, the lithium-ion battery will play increasingly important roles in the area of power tools as the European Union and China are further limiting the use of nickel-cadmium batteries.
- The lithium-ion power battery is also finding application in mine-lamp, UPS, and communications.

Table 1. Production of Secondary Batteries in 2002–2006 in China.

Year	Lead-Acid Battery Production (million unit)	Nickel-Metal Hydride Battery Production (million unit cells)	Nickel-Metal Hydride Battery Export (million unit cells)	Nickel-Cadmium Battery (million unit cells)
2002	500	-	-	-
2003	600	-	-	-
2004	750	-	-	-
2005	900	960	870	350
2006	1,005	1,100	960	430

The market analysis of positive materials for the lithium-ion power battery:

- $LiCoO_2$ is the main positive materials for the lithium-ion battery. Since 1990, the lithium-ion battery has been commercialized in many developed countries, such as Japan, the United States, France, and Germany. In China, the lithium- ion battery had been commercialized by the end of 20th century. The lithium-ion battery with $LiCoO_2$ is being developed with the goals of longer life, higher capacity, and higher safety. The Chinese $LiCoO_2$ industry has to make significant efforts in research and development of new products to be competitive and keep development sustainable.
- The $LiMn_2O_4$ lithium-ion battery is finding applications in portable electronics, communication, military equipment, and transportation. The batteries could be used as energy-storage devices in applications for exploring wind and solar energy.
- The $LiFePO_4$ lithium-ion battery offers good safety and performance in a high-temperature environment, as well as good capacity and low cost (which is only one-

quarter that of LiCoO$_2$). The LiFePO$_4$ lithium-ion battery is finding application in energy-storage devices for solar- and wind-generator systems, UPS, power tools, electric vehicles, medical equipment, toys (remote electric toy planes, toy vehicles, and toy boats, for example), and others.
- For the lithium-ion power battery, LiMn$_2$O$_4$, LiFePO$_4$, and LiMn$_x$Ni$_y$Co$_{1-x-y}$O$_2$ will share the market as positive materials in the coming three years. After three years, LiFePO$_4$ will occupy a much bigger market. In three years, the demand for LiFePO$_4$ is estimated to be over 10,000 tons/year.
- So far, the main competition for LiFePO$_4$ in China comes from Valence Technology, Inc. (USA); A123 (USA); and Tianjin STL Energy. However, most of the LiFePO$_4$ companies (e.g., Huannan Reshine and Pulead Technology Industry Co., Ltd.) have stability (consistency) issues between batches of production.

The sources distribution of raw materials for the lithium-ion power battery are summarized below:

- Lithium sources: Lithium ore deposition in China is mainly distributed in Sichuan (51.1%), Jiangxi (29.4%), Hunan (15.3%), and Xinjiang (3%), and the rest (only 1.2%) is distributed in Henan, Fujian, and Shanxi provinces. Lithium sources from salt lakes are mainly distributed in Qinghai, Tibet, and Hubei; among them, 80% of lithium from salt lakes is in Qinghai.
- Cobalt sources: Sources of cobalt in China are limited; 30% of cobalt is from Gansu province.

Information about selected lithium-ion battery companies is summarized below:

Table 2. Growth of Demand for Lithium-Ion Power Batteries in 2001–2007 in the Global Market.

	Year						
	2001	2002	2003	2004	2005	2006	2007
Growth in Demand (%)	8.3	9.2	10.1	11.6	12.7	13.6	15.3

Table 3. Production and On-Road Number of Automobiles, Motorcycles, and Electric Bikes in 2001–2007 in China.

Year	Automobile Production (10^6)	On-Road Automobiles (10^6)	Motorcycle Production (10^6)	On-Road Motorcycles (10^6)	Production of Electric Bicycles (10^6)	On-Road Electric Bicycles (10^3)
2001	2.344	18.02	9.96	47.6	0.4	-
2002	3.251	20.53	11.5	51.0	1.6	-
2003	4.444	23.83	15.0	60.0	4.0	-
2004	5.074	27.42	14.75	67.5	6.75	13.0
2005	5.708	35.0	17.24	76.3	12.11	23.2
2006	7.280	41.0	21.45	83.5	19.5	37.5
2007 (Prediction)	8.50	47.5	25.4	94.0	-	-

Table 4. Growth in Demand for Electric Bikes, Power Batteries for Electric Bikes, Production of Batteries for Electric Bikes, and Exportation of Power Tools in 2001–2007 in China.

Year	Growth in Demand for Electric Bicycles (%)	Growth in Demand for Electric Bicycle Batteries (%)	Growth in Demand for Electric Bicycle Batteries (%)	Growth in Exportation of Power Tools (%)
2001	36.5	41	32.3	10.2
2002	300	282	216.4	11.7
2003	150	163	137.5	12.8
2004	68.8	65	58	13.6
2005	79.4	81	76	15.9
2006	61	59	67	17.6
2007 (prediction)	59	-	-	-
2008 (prediction)	63	-	-	-

1. Suzhou Phylion Battery Co., Ltd.

Suzhou Phylion is a battery technology corporation set up by Legend Capital Co., Ltd.; Institute Biophysics Chinese Academy of Sciences; and Chengdu Diao Group. The company has 82 million RMB in capital, over 400 employees, and production capacity of 36 million A•h/yr.

2. Thunder Sky Battery Limited

Thunder Sky Battery Limited (TS) is a high-tech manufacturer that is the first company in the world to successfully replace Polyvinylidene Fluoride (PVDF) by using solvent binder in the production of rechargeable lithium-ion batteries with high capacity and high power. Since its foundation in 1998, TS has gained a number of patents in over 26 countries and product areas for its original lithium-ion power battery technology, which is a patented solid-state lithium-ion power battery based on a liquid lithium-ion battery and solid polymer lithium-ion battery.

TS considers its lithium-ion battery with high power, high capacity, and high voltage not only the ideal energy replacement for fuel and the perfect traction energy source for environmental protection transportation tools, but also the optimum power for mobile energy, solar energy, wind energy, and other multi-power systems in military use. The TS lithium-ion battery is authorized by the U.S. Department of Transportation as a safe battery and is therefore allowed to be shipped worldwide. TS is the only company to be awarded this certification from the United States. Known as "non-explosive battery" since its first entry into market in 1998, the TS battery has been used as a main source of power in tourist submarines and regular submarines and has been extremely successful in deep-sea applications of high-power, high-capacity lithium-ion batteries. TS battery has an excellent reputation in Europe, Asia, Africa, and North America.

With its fast expansion, TS moved to its new base in September 2006, and it will carry out complete automatic production at large scales, which will increase the yearly output capacity from 150,000,000 A·h to 1,500,000,000 A·h. To meet the increasing demands of overseas customers, TS will build a joint-venture factory in the United States and Finland that will work as the production and service center of TS in Europe and North America. At present, an electric bus using TS lithium-ion batteries has started operation in North America and Finland. Local Shenzhen government has approved the use of the TS battery in the electric bus, and in the near future, the TS battery will be used in electric vehicles worldwide, once the United States and Finland start the production of EV buses.

By applying its philosophy to "obtain one percent progress every day," TS will make the needed improvements in quality and service for its loyal customers to realize its goal to "be beneficial to people and be profitable to enterprise."

3. Huanyu Power Source Co., Ltd.

With its headquarters in Xinxiang, Henan, Huanyu Power Source Co., Ltd., began its main business of R&D, manufacturing, and distribution of secondary batteries in 1982. Huanyu has more than 10,000 employees, among them 400 who are dedicated to research.

Huanyu has established a high-tech industrial park engaged in the production of four series of batteries (namely, nickel cadmium, nickel hydride, lithium-ion, and lead acid batteries) in more than 100 cell phone models, with a daily capacity in excess of 2 million secondary batteries in these models. The industrial park is the largest production base of secondary batteries with the widest range of models in China.

4. Qing Dao Aucma Newpower Technology Co., Ltd.

Qing Dao Aucma New Power Technology Co., Ltd., was established and is financed by the main shareholder, Qingdao Aucma Co., Ltd. This high-tech company develops new energy products in the field of environmental protection. With registered capital of ¥90 million and total investment of over ¥200 million, its main products have been listed in the national project of "Two Hi-Tech and One Refinement" (industrializing high-tech products, remolding the traditional enterprise with high-tech products, and refining the product structure). With the support of the state-class enterprise technical center and the state high-tech industrialization policy, Qingdao Aucma is industrializing and developing high technology while maintaining its own intellectual property rights. Qingdao Aucma is devoted to developing and producing the key parts needed for the environmental protection of an energy lithium-ion battery cell, including prismatic, cylindrical and steel shell, aluminum shell, and soft-package types of batteries, which can be widely used in the mobile telephone, battery, laptop computer, portable machine battery, electric toy, and power battery industries, among others. The company can also design and produce special types of battery cells as required. The company has successfully passed ISO9001 quality management system certification and is managed strictly according to ISO9001 standards. The products have CE certification. According to appraisals by outside experts, all performance data of Qingdao Aucma's products are world-class. The company's lithium-ion battery cell has the superior characteristics of high capacity, long cycle life (700–800 times), and low self-discharge rate, among other attributes. Aucma New Power will uphold the principles of Aucma Group Good, which include ensuring cooperation and mutual development, presenting reliable products with reliable moral quality, and striving for the great development of the Chinese energy industry in the new century.

5. Wuhan Lixun Power Corp., Ltd.

Wuhan Lixun Power Corp., Ltd., is a new, major, national-level and high-technology enterprise established in May 1993. It is located in the Great Wall Innovation Scientific and Technological Garden of Wuhan East Lake High-tech Developing Zone. This location is characterized by a high intellectual concentration in the Optical Valley and includes a 30,000-m^2 workshop and 24 self-design lithium battery production lines that produce 20,000 pieces annually (the output of Li/MnO_2 button batteries ranks the third in the world); the output of lithium batteries is 1,800 pieces and 80,000 sets of programmed testers. Wuhan Lixun Power Corp., Ltd., has demonstrated itself to be a professional lithium battery base for those engaged in the study, production, and trade field of Li/MnO_2, $Li/SOCl_2$, and lithium-ion batteries, as well as a programmed battery tester.

At the end of 2005, a new workshop located in the Great Wall Innovation Scientific and Technological Garden of Wuhan East Lake High-tech Developing Zone was established, occupying an area of 930 m^2, for the production of charged lithium batteries. It was put into production in February 2006. The new workshop has not only increased production capability and satisfied the increasing market demand, but it also has built a solid foundation for the construction of large-scale, industrialized production of lithium batteries.

The product manufactured by the corporation is listed as one part of the state council Torch Plan, an industry test project of State Economic and Trade Commission, a major new product special project of Hubei Province, and a major new product development project of Wuhan. The company owns a state patent and a utility model of more than 20 items in terms of series lithium- ion batteries. The company has been listed as a key electromechanical product export enterprise by Hubei Province; the products manufactured by the company are exported to Europe, North America, Southeast Asia, Taiwan, and Hong Kong.

6. Tianjin Lishen Battery Joint-Stock Co., Ltd.

Tianjin Lishen Battery Joint-Stock Co., Ltd., which is situated in Tianjin Huayuan Hi-Tech Industry Park, occupying a total area of 85,000 m^2, has a registered capital of 600 million RMB, and its total investment has reached 1.5 billion RMB. Lishen has imported advanced automatic equipment from abroad. As of 2004, it had already achieved an annual production capacity of 200 million cells, which consist of more than 100 specifications, ranging from the cylindrical lithium-ion battery (LIB) to the prismatic LIB, laminated LIB, and lithium-ion polymer (LIP). Relying on its independent intellectual property rights, with innovation-oriented organizational and supporting policies, Lishen has become one of the largest lithium-ion battery manufacturers and possesses the most advanced technology in

China. The world-renowned *Forbes* magazine, Chinese version, ranked Lishen at the 8th position on the "List of Most Potential Enterprises" in China 2006.

Since its founding, Lishen has been dedicated to the concepts of "Developing science and technology by the people and for the people; Focusing on management; putting quality the top priority; Pursuing verity and innovation." The quality and performance of Lishen cells are world- class. Meanwhile, Lishen has proudly obtained the certification of ISO9001:2000, CE, UL, and ISO14001. The successes associated with these third-party certifications have paved a way for Lishen to get into the international and domestic markets. So far, Lishen has supplied batteries to multi-national corporations (like Motorola ESG and Samsung). At present, Lishen has set up branches in North America, Europe, Korea, and Hong Kong, establishing a powerful worldwide marketing network.

In the industrialization process, Lishen is aware of the importance of independent intellectual property rights for core technology. Therefore, it has always attached great importance to technology research. Investment on R&D is increasing. Lishen has already set up a Postdoctoral Workstation and a National Technology Center. In 2005, it established a world-class Safety Test Center.

7. CITIC GUOAN Mengguli Corporation (MGL)

Located in Beijing Zhongguancun Science Park, CITIC GUOAN Mengguli Corporation (MGL) is engaged in the research, development, and production of new composite metal oxide materials and high-energy-density lithium-ion secondary batteries. MGL is invested in primarily by CITIC GUOAN Group, a wholly owned subsidiary of China Zhongxin Group (CITIC). The CITIC GUOAN Group has operations in industries including information technology, new materials, mineral resource surveying, tourism, and real estate. Ratified by Deng Xiaoping, CITIC was founded in October 1979 by Rong Yiren, former Vice Chairman of the Peoples Republic of China. After more than 20 years of growth, CITIC is now a large-scale international enterprise group with total assets of ¥700 billion.

MGL is China's largest manufacturer of the lithium-ion cathode material $LiCoO_2$ and is first in line to market the new cathode materials $LiMn_2O_4$ and $LiCo_{0.2}Ni_{0.8}O_2$. Being quality-oriented, MGL has been certified to both of China's New and High-Tech Enterprise standards and to ISO9001:2000. MGL's unique synthesis method simply and efficiently produces cathode materials of superior electrochemical performance and reliability in an environmentally friendly way. Since incorporation, MGL has smashed the monopoly of China's lithium-ion battery cathode materials market held by foreign manufacturers and now stands at the forefront of the industry. Besides cathode materials, MGL also produces high-capacity, high-energy-density lithium-ion secondary batteries for power and energy storage, with capacities ranging from several ampere-hours to several hundred ampere-hours. As China's first and only power battery manufacturer, MGL is now setting the global pace by presenting high-capacity lithium-ion secondary batteries, which have been applied successfully to Beijing's fleet of trial electric buses.

To ensure sustainable and steady development, MGL has built up a modern R&D department in Beijing. Through the combined efforts of MGL staff, MGL is able to contribute more and more to social progress and development.

$LiCoO_2$, $LiMn_2O_4$, and $LiCo_{0.2}Ni_{0.8}O_2$ are the core products of MGL's Materials section. These oxide materials are indispensable to high-voltage (4 V) and high-energy-density lithium-ion secondary batteries. Over the past 10 years, lithium-ion secondary batteries have taken the place of NiMH and NiCd secondary batteries in a wide variety of applications, including mobile phones and laptop computers.

Instead of the commonly used solid-state synthesis method, MGL has adopted a unique method to synthesize materials that is a highly efficient, simple process with zero emissions and low energy consumption. Feedback from lithium-ion battery manufacturers in China and abroad indicates that MGL's battery cathode materials are excellent and steady electrochemical performers. MGL is China's largest lithium-ion cathode materials manufacturer and holds a competitive and leading position in the global market. However, despite the rapid development of lithium-ion batteries over the last ten years, limited cobalt resources and poor thermal stability of $LiCoO_2$ could restrain the practical scope of lithium-ion batteries.

With this in mind, and with the steady support of Chinese State and local governments, MGL has focused on developing new lithium-ion battery cathode materials, notably spinel $LiMn_2O_4$ and layered $LiCo_{0.2}Ni_{0.8}O_2$. Experiments have proven that the superior thermal stability and steady charge-discharge performances of $LiMn_2O_4$ and $LiCo_{0.2}Ni_{0.8}O_2$ qualify them as suitable cathode materials for various types of lithium-ion batteries. Recently, $LiMn_2O_4$-based and $LiCo_{0.2}Ni_{0.8}O_2$-based lithium-ion batteries have been applied in a variety of energy-saving and environmentally friendly industries. China is poor in cobalt deposits but boasts large deposits of manganese and nickel. MGL is dedicated to developing China's lithium-ion battery materials industry based on its pioneering synthesis method and to contributing more to the development of the communications industry and new energy and environmentally friendly industries.

Differing from NiMH and NiCd secondary batteries, lithium-ion secondary batteries are a ready source of high voltage (4 V), small size, and lightweight power. Offering greater flexibility under different temperatures, they suffer no memory effects and create minimal pollution. At present, lithium-ion secondary batteries have an energy density two to three times higher than that of lead acid batteries and around twice as high as NiMH and NiCd batteries. Small lithium-ion batteries have been widely applied to small high-end electronic devices, such as mobile phones and portable computers. With the progress of chemistry and materials science, commercialized lithium-ion batteries will offer greater improvements in performance and an expansion of their applications. These developments are due to the important role played by the physical-chemical properties of cathode materials, separators, and electrolytes in the reliability of lithium-ion batteries; in addition, carbon has realized only one-tenth of its theoretical capacity as an anode material. Therefore, new organic, inorganic, and metallic compounds will improve the physical-chemical performances of lithium-ion batteries and expand their applications, while solid and inorganic electrolyte will drastically improve the reliability and safety of lithium-ion batteries.

Recently, MGL has independently developed high-capacity lithium-ion batteries and successfully applied them to Beijing's trial fleet of electric buses. Experiments indicate that new lithium-ion batteries are expected to speed up the industrialization of electric vehicles and show great potential in such applications as mobile communications, nighttime power storage, wind- and solar-power storage, backup emergency power, backup power for vehicles, and portable power.

As the agent of many vital state and provincial research projects, MGL has focused on the R&D of battery technologies when developing its materials section. During the past 10 years, lithium- ion power batteries for electric vehicles have been a global focus, and the aim is now to industrialize electric vehicles. MGL's high-capacity lithium-ion batteries will ease pressing energy consumption concerns for urban transportation and solve issues of pollution. This will pave the road for China's development of an innovative automobile industry.

8. TCL Hyperpower Batteries, Inc.

Founded in 1999,TCL Hyperpower Batteries, Inc., is a subsidiary company of TCL Corporation. It specializes in the design and manufacturing of high-energy lithium-ion batteries. The company has been investing its resources in the development of core technologies and processes by building a strong team of research and development (R&D), quality control, Professional Engineering (P.E.), and management, and TCL is constantly upgrading its manufacturing facilities. TCL has become a leading designer, manufacturer, and supplier of lithium-ion battery cells in China.

The company emphasizes the development of intellectual property and possesses its own core technology patents. It has ISO9001/2000, ISO14001, UL, CE, and RoHS certification. Six sigma management is widely employed for improving the processes and refining TCL's expertise toward zero-defect production.

TCL's leading technology, process-engineering capability, manufacturing infrastructure, and management strength enable it to provide a unique scope of total battery solutions and to add value to its customers.

9. China Powerel Battery Co., Ltd.

China Powerel Battery Co., Ltd., a high-tech company specializing in R&D, manufacture, and sales of rechargeable lithium-ion polymer battery and safety power supply, was established in April 2003. The headquarters is in Beijing Zhongguancun Science & Technology Park.

China Powerel has high-throughput and advanced technology in polymer lithium-ion batteries and implements a scientific quality control system of ISO9001 and Lean Manufacturing. The company's products have been certified by the organizations of CE, UL, and SGS. The testing center is self-contained and has the capability to inspect electrical property and security according to international and industry standards. China Powerel forged a strategic alliance with enterprises that engage in new materials R&D and in the design of protection circuits. An assembly line for battery configuration has been set up to provide integrated power solution to its customers. The daily throughput is 50,000 A•h, and the company is fully capable of offering fast, effective service to its customers.

The company's products range from tens of milliampere hours to dozens of ampere hours; the shape and dimension can be varied to meet customer requirements. The discharging rate of the power battery is up to 15°C. The application fields include mobile products, digital products, consumer electronic products, electric tools, electric bicycles, electric cars, aero-models, toys, miner's lamps, pharos, medical devices, and military equipments.

Its sole subsidiary company — China Power Battery Technology Co., Ltd. — was established in September 2005 and is engaged in R&D and manufacturing power batteries with features of high capacity, high power, and high security. Applications include electrical bicycles, electrical cars, electrical tools, and aero models.

The company's goal is to develop into a integrated enterprise with core competency and international fame in the field of the rechargeable lithium-ion batteries, taking system innovation as a guidance, technology innovation as a base, service innovation as a means, and capital expansion as the impetus.

10. Zhejiang Xinghai Energy Technology Co., Ltd.

Zhejiang Xinghai Energy Technology Co., Ltd., is located in Taihu Road Eco & Tec Development zone, Changxing County Zhejiang, China. It specializes in producing cylindrical lithium-ion batteries, motive lithium-ion batteries, and electric-bicycles powered by lithium-ion batteries. At present, annual production capacity is 5 million cylindrical lithium-ion cells, 100,000 packs of motive lithium-ion batteries, and 120,000 of e-bicycles driven by lithium-ion batteries. These batteries are widely used in various areas, including mobile phone batteries, power tools batteries, wireless earphone batteries, blue tooth technology, computers, digital cameras, video cameras, and electric vehicles, for example.

The company's e-bicycle with lithium-ion batteries adopts batteries with lightweight metal in positive materials. Such batteries are favored because they offer safety, long cycle life, good discharging ability at high load, resist high temperature, offer stable charging and discharging performance, and are environmentally friendly.

11. Shanxi Guangyu Power Sources Co., Ltd.

Shanxi Guangyu Power Sources Co., Ltd., was established in 1988. It is a high-technology company that focuses on developing and manufacturing powerful LED, LED lighting products, lithium battery LED cap lamps, and power lithium battery packs. The company's lithium batteries and powerful LED lighting products have already become world famous. Shanxi Guangyu Power Sources Co., Ltd., is good at powerful LED encapsulation and application. As a brand, Jiebell products have already been used in different fields, like roads, tunnels, underground mines, offices, and workshops. They are both for indoor lighting and outdoor lighting. Power lithium battery packs are used widely for electric bicycles.

With many years of development, the quality of the products and company management of Shanxi Guangyu Power Sources Co., Ltd., have entered into a new stage. The company has passed the Quality Management System Certificate ISO9001 and Environmental Management System Certificate ISO 14000. It also gained many distinctions:

- "The Key Enterprise of Creating Famous Brand in China" by the Ministry of Agriculture of the People's Republic of China

- "Contract Abiding and Trustworthy Enterprise" and "The Enterprise of Creating Famous Brand in Shanxi" by the State Administration for Industry and Commerce
- "High Technology Enterprise" the by Department of Science and Technology, Shanxi Province
- "AAA Class Trustworthy Enterprise" recognized by the Agriculture Bank of China, Shanxi Branch
- Shanxi Guangyu Power Sources Co., Ltd., recognized as a technology center of Shanxi Province
- The brand "Jiebell" is recognized as a famous brand by China's Administration of Industry and Commerce Bureau in Shanxi Province
- LED civil lights are considered to be power-saving architectural products by Shanxi Construction Office and have been awarded the certificate of New Technology in Chongqing Civicism by Chongqing
- Company Executive Officer (CEO) serves as Vice-Chairman Member of Solar Energy Photovoltaic Lighting Committee in China Illuminating Engineering Society
- CEO serves as A Commissioner of Semiconductor Illumination in China Association of Lighting Industry
- Member of "Lighting Africa," which was initiated by the World Bank Group (International Finance Corporation-World Bank, or IFC-WB)
- CEO serves as Editor of LED Road Lighting Standard in China

To stir the development of the LED industry, Shanxi Guangyu Power Sources Co., Ltd., invested RMB 500 million for Shanxi Guangyu Photoelectron Industry Garden, which occupies 300,000 m^2 of manufacturing space. The company used the most advanced encapsulation product lines for high-power LED and application lines for LED lighting products.

Shanxi Guangyu Photoelectron Industry Garden is expected to become the largest production base for single powerful LED and LED lighting products and will be manufactured with advanced technology and have the highest quality. The production value is estimated to be up to RMB 10 billion. The products are expected to contribute significantly to energy-saving efforts around the world.

12. Tianjin Hang Li Yuan Technologies, Inc.

With the establishment of cooperation between Tianjin DaMing Vehicle Industry, Ltd., and the Institute of Chemical Physics power supply of the Chinese space group, Tianjin Hang Li Yuan Technologies, Inc., was established with an investment of ¥300 million in 2004.

On the basis of technology of a power supply plant for space flight, a vanguard team was formed to study lithium-ion batteries. Hang Li Yuan insists on scientific management and technological innovations to offer the best products for clients and careful attention to detail. The company adheres to the ISO9000 quality system to strengthen quality management and remain competitive. The company also initiated the "5 Star" system to award employes with high production to improve staff quality and foster a good and healthy company image.

The company primarily produces and deals with drive lithium-ion rechargeable batteries. Production has reached 10,000 A·h each day, at a capacity of 5,000 A·h daily. The present production system can be classified into four categories: $LiCoO_2$ system, $LiMn_2O_4$ system, $LiFePO_4$ system, and three elementary material systems. The production model includes 3-, 3.5-, 4.5-, 5-, 7-, and 10-A·h units. As a power supply, the product has wide applications, including electric bikes, electric autobikes, electric cars, electric implements, mobile telephones, laptop computers, digital cameras, electronic apparatus, and medical instruments. At the same time, the company can also produce different styles of batteries to meet customer needs.

13. Suzhou Dinet Energy Tech Co., Ltd.

Suzhou Dinet Energy Tech Co., Ltd., was founded on November 18, 2004. It focuses on high- energy green power: lithium-ion batteries, solar cells, and power sources.

14. Shuang Yi Li (Tianjin) New Energy Co., Ltd.

Shuang Yi Li (Tianjin) New Energy Co., Ltd., was founded by Japan DKS, Tianjin Yiqin Group, and Japan ENAX. By applying the technology of polymers from DKS and lithium-ion production from ENAX, it produces lithium-ion polymer batteries and power batteries for electric bikes, power tools, UPS, electric vehicles, and hybrid electric vehicles.

15. Shenzhen Xingke Professional Li-ion Battery Co., Ltd.

Shenzhen Xingke Professional Li-ion Battery Co., Ltd. (XKTD), was invested in and founded by the Shenzhen Rongxing Group and is a high-tech enterprise engaging in the R&D, manufacturing, and marketing of lithium-ion polymer batteries.

XKTD produces a wide range of lithium poly batteries to meet the needs of various electronic devices, including Bluetooths, MP3/MP4, cellular mobile phones, DVDs, palmtops, notebooks, computers, personal data assistant (PDA), digital cameras, camcorders, electronic toys, and tools. XKTD's flexible manufacturing capabilities and the use of latest technology enables custom designs in a timely and cost-effective way.

XKTD highly values innovative products, reliability, and quality, as well as comprehensive customer services, as the means of maintaining its competitiveness and directing the development in today's marketplace. Its R&D department consists of a team of more than 20 engineers, including electrics matching and testing senior engineers, PCB (printed circuit board)-designing senior engineers, and battery decoding experts. With its professional knowledge and abundant experience, the company says that it provides customers with reliable, steady, and trustworthy products.

In addition, XKTD employs rigorous quality control procedures (e.g., IQC, IPQC, OQC, FQC, QA) associated with a modern management system and has the approvals of ISO9001:2000, CE, UL, and RoHS. The company maintains quality assurance throughout each of its research and production links, enabling it to supply reliable products to customers.

16. Jiangxi Meiya Energy Co., Ltd.

Jiangxi Meiya Energy Co., Ltd., relies on the R&D of Nanchang University. Its core products are lithium-ion power battery and packs. Its production is 10 million A•h/yr.

17. Tianjin Blue Sky Double-cycle Tech. Co., Ltd.

Tianjin Blue Sky Double-cycle Tech. Co., Ltd., founded by China Electronic Tech Group Corporation and Tianjin Metallurgical Group Co. Ltd., is a high-tech enterprise.

With a registered capital of 93 million RMB, it specializes in the research, development, and marketing of lithium-ion batteries and electric-bikes. The company with a cogent technical capacity undertakes the electric vehicle project of the National 863 Program and a number of Tianjin's key projects; its products include lithium-ion batteries, efficient electric machines, controllers, and electric-bikes and related parts.

Of its total 50 researchers, more than 10 are professor-level senior engineers, and the engineering staff makes up 45% of its total employees.

18. Hunan Haixing High-tech Power Battery Co., Ltd.

Hunan Haixing High-tech Power Battery Co., Ltd., is one of the leading corporations professionally developing lithium-ion power batteries in China. The company is supported by advanced technology and equipment, as well as an excellent team of designers and researchers, who consist of professionals specializing in lithium-ion batteries from abroad, and experienced technicians.

The company was established in May 2006, with a total investment of 50 million. The company is located in Hunan Taishang Developing Zone in Wang Cheng County. The company has been well managed and has developed a variety of batteries with reliable performance and safety, which led to ISO9000 qualification and CE Certification, as well as other credentials by the Chinese Authority.

With different capacities and dimensions, Hunan Haixing offers customers a variety of choices; for example, it can provide batteries with capacities ranging from 1,600 mA•h to 50 A•h. Products can be applied in electric tools, notebooks, electric motorcycles, electric vehicles, miner's lamps, and portable equipment. The company can also provide other types of lithium-ion batteries of different capacities and dimensions, depending on customer needs.

19. DLG Battery (Shenzhen) Co., Ltd.

Founded in 2001, DLG Battery Co., Ltd., specializes in the research, manufacture, and marketing of rechargeable batteries and relevant products, such as lithium-ion, nickel-metal hydride, nickelcadmium, battery chargers, and flashlights. With factories located in Shanghai and Jiangmen, DLG Battery provides OEM (original equipment manufacturers) manufacturing services and has promoted successfully three brands of batteries and relevant products in China's consumer markets, namely "CISHIDAI, " "DELANG," and "DLG."

DLG Battery (Shenzhen) Co., Ltd., specializes in conducting research, manufacturing, and marketing batteries and related products, including nickel-metal hydride, nickel cadmium, and lithium-ion batteries, and battery chargers. With its advanced manufacturing equipments, scientific operational believes, well-trained personnel, and quality products, DLG Battery has created an image of "battery expert" in the industry and developed quickly over the past few years. At present, DLG Battery has more than 100 major agents in mainland China, and its sophisticated distribution networks cover almost every province. Meanwhile, its products are well received in South and North America, Europe, Asia, and Oceania.

DLG Battery (Shanghai) Co. Ltd. is one of the top producers of cylindrical rechargeable lithium- ion batteries in China. It was founded in October 2001 and located in one of China's most dynamic regions — Shanghai — occupying a total area of 36,000 m^2 and a total construction area of 10,000 m^2. DLG Shanghai specializes in researching, manufacturing, and marketing rechargeable batteries, especially high-rate cylindrical lithium-ion batteries and polymer batteries, and its products are widely used in electrical tools, digital cameras, video cameras, portable DVD players, notebook PCs, MP4 players, and mobile communication equipment. DLG Shanghai is a qualified supplier to some world-renowned companies, such as LG Electronics. In June 2006, DLG Shanghai joined with Peak Energy Solutions, Inc., an American company with advanced technology in large format battery manufacturing, to develop power Lithium Iron Phosphate (LIP) batteries.

The R&D Center of DLG Shanghai consists of more than 30 senior engineers with PhD or Master's degrees and is technically backed up by such research institutions as the University of Wollongong in Australia, Shanghai Jiaotong University, and Shanghai University, among others.

DLG Shanghai has been certified with "ISO9001:2000" Quality Management System in May 2004, and it acquired CE and UL certificates and passed ROHS tests. In 2004, DLG Shanghai was ranked "New and High-Tech Enterprise of Shanghai."

Over the years, DLG has made considerable progresses in promoting the "DLG" brand in overseas markets. On the basis of the principles of equality and mutual benefit, the company has cooperated well with its partners worldwide and has built extensive distribution networks in Japan, Republic of Korea, Malaysia, Germany, Australia, New Zealand, Brazil, and others. DLG says that it welcomes partners worldwide to join it in exploring the world market and sharing in its continued growth and success.

INDEX

#

20th century, 159, 217
21st century, 105

A

abuse, 9, 20, 56, 66, 168, 179
access, 106
accessibility, 96
accounting, 83, 89, 126, 134
acid, 2, 5, 14, 21, 24, 26, 29, 30, 35, 50, 51, 54, 77, 79, 81, 82, 83, 84, 86, 92, 94, 97, 99, 100, 107, 112, 116, 117, 120, 124, 132, 133, 134, 137, 138, 140, 142, 151, 157, 159, 160, 165, 167, 206, 212, 213, 215, 216, 220, 223
adaptability, 97
additives, 16, 25, 130, 131
adults, 110
advancements, vii
Africa, 92, 219, 226
agencies, 3, 35, 79, 118
air pollutants, 104
air quality, 90, 102, 117
airports, 114
alternative energy, 24, 82, 118, 165, 203
aluminium, 2
ambient air, 13
amplitude, 12
annual rate, 22, 81, 101
appraisals, 220
arbitration, 35
Asia, vii, 3, 9, 79, 82, 86, 105, 118, 172, 219, 229
assessment, 9, 86
assets, 29, 34, 136, 144, 145, 151, 222
atmosphere, 83, 95, 127
atoms, 16
authentication, 142
authorities, 96, 97, 138
authority, 55, 155, 167
automation, 7, 48, 49, 122
Automobile, 40, 42, 43, 44, 57, 64, 65, 77, 98, 104, 108, 111, 146, 174, 175, 218
automobile parts, 151
automobiles, 48, 50, 51, 81, 93, 94, 98, 100, 101, 117, 155, 156, 157, 160, 203, 216
automotive application, 9, 10, 16
automotive applications, 9, 10, 16
avoidance, 20
awareness, 111

B

backlash, 100
balance of payments, 87
ban, 80, 83, 89, 90, 91, 120
Bangladesh, 107
banks, 46, 134
base, 24, 51, 53, 97, 122, 135, 140, 141, 160, 161, 162, 164, 220, 221, 225, 226
basic research, 6, 35, 45
battery option, vii, 3
bauxite, 145
BBB, 144
Beijing, 1, 6, 7, 8, 25, 26, 28, 29, 30, 32, 33, 34, 36, 40, 43, 44, 45, 46, 47, 48, 49, 55, 56, 57, 58, 59, 60, 63, 64, 65, 72, 73, 74, 75, 79, 81, 83, 84, 94, 98, 99, 101, 102, 103, 119, 120, 122, 131, 132, 134, 135, 136, 137, 161, 164, 171, 173, 174, 178, 179, 216, 222, 223, 224
Belgium, 80, 92, 171
benchmarking, vii, 5, 8, 32, 34, 39, 42, 44, 56, 79, 85, 168
benefits, 80, 89, 96, 102, 110, 165
bias, 166
biomass, 52, 53, 123

Bluetooth, 144
Brazil, 229
business model, 116
businesses, 81, 99, 111, 138, 211, 215
buyer, 109
buyers, 7, 55, 111

C

C++, 58
cadmium, 2, 22, 48, 54, 75, 84, 121, 216, 217, 220, 229
caliber, 34
calibration, 34
Cambodia, 107
camcorders, vii, 3, 7, 27, 45, 47, 118, 140, 146, 227
candidates, 29, 137
carbon, 10, 16, 20, 24, 30, 36, 75, 99, 110, 117, 123, 129, 223
carbon dioxide, 75, 99
carbon materials, 129
carbon monoxide, 75, 117
casting, 115
catalyst, 82, 105
cathode materials, 5, 25, 29, 30, 32, 84, 123, 127, 128, 136, 155, 222, 223
cell phones, vii, 3, 8, 21, 23, 45, 49, 83, 84, 118, 120, 122, 123, 125, 126, 140, 142, 157
cell size, 18, 25, 40
Census, 133
ceramic, 52
certificate, 49, 226
certification, 6, 29, 34, 37, 138, 139, 142, 143, 144, 147, 162, 163, 219, 220, 222, 224
challenges, 16, 85, 97, 122
chemical, 6, 10, 16, 20, 30, 34, 35, 52, 117, 130, 131, 223
chemical properties, 30, 223
chemical reactions, 20
chemical reactivity, 16
chemical stability, 10, 131
chemicals, 20, 52
Chicago, 212
children, 161
Chile, 133
Chinese government, 4, 5, 7, 8, 21, 29, 37, 46, 48, 49, 53, 55, 56, 83, 84, 120, 121, 133, 165
Chinese policies, regulations, vii, 79, 85
cities, 49, 80, 81, 83, 84, 88, 89, 92, 93, 94, 95, 96, 98, 99, 101, 102, 103, 105, 108, 110, 111, 132, 162, 164, 165
citizens, 83, 87, 110, 132

City, 24, 34, 42, 50, 51, 58, 59, 60, 107, 131, 142, 144, 146, 161, 172, 176, 178, 201, 202, 208, 214
classification, 91, 96, 109
clean energy, 52, 151
clients, 35, 226
climate, 13
climates, 4, 13, 14, 15
CO2, 75, 99, 107, 117
coal, 54, 90, 91
cobalt, 2, 5, 10, 11, 24, 30, 33, 218, 223
coke, 10
collaboration, 22, 98
color, 113
combustion, 90
commerce, 205
commercial, 21, 24, 99, 119, 140
communication, 37, 110, 117, 128, 129, 159, 217
communities, 103, 211
community, 111
compatibility, 129, 132
competition, 23, 99, 101, 104, 125, 126, 129, 164, 165, 217, 218
competitive advantage, 8, 85
competitiveness, 10, 118, 227
competitors, 100, 159
complement, 159
compliance, 108
composites, 12, 45
compounds, 223
compression, 44
computer, 75, 146, 157, 220
computer-aided engineering (CAE), 157
conditioning, 13, 14, 53
conductivity, 10, 11, 12, 13, 124, 128, 131
confidentiality, 71, 199
configuration, 31, 224
Congress, 53, 61, 96, 99, 109
construction, 15, 16, 17, 44, 53, 94, 221, 229
consulting, 212
consumer markets, 229
consumers, 7, 55, 88, 95, 108, 112, 124, 126, 132, 165
consumption, 9, 86, 88, 123, 128, 129, 136
Continental, 99
cooking, 115
cooling, 3, 4, 12, 13, 14, 15, 35, 114
cooperation, 39, 42, 45, 57, 147, 221, 226
coordination, 53
copper, 10, 19, 145
cost, 3, 4, 5, 7, 9, 10, 11, 12, 13, 18, 19, 20, 23, 25, 29, 32, 33, 38, 48, 49, 51, 53, 56, 57, 66, 71, 80, 81, 82, 89, 95, 96, 97, 98, 102, 105, 109, 111, 113, 115, 116, 117, 118, 122, 123, 124, 125, 126,

Index

128, 129, 132, 151, 156, 158, 159, 165, 168, 179, 199, 217, 227
cost constraints, 10
covering, 45, 164
CPU, 75, 100
credentials, 228
crude oil, 87
crystal structure, 11
crystalline, 54
culture, 80, 92, 101, 145
currency, 77, 86
customer service, 227
customers, 8, 71, 83, 85, 88, 120, 132, 142, 144, 147, 150, 162, 200, 220, 224, 227, 228
cycles, 15, 17, 18, 42, 46, 67, 83, 112, 116, 126, 127, 143, 144, 145, 155, 180, 181, 203, 207
cycling, 10, 16, 54, 102, 103
Cyprus, 133
Czech Republic, 61

D

decibel, 78
decoding, 227
decomposition, 16
deficit, 87
deformation, 42
degradation, 4, 19, 28, 33, 137
degradation process, 4, 19
Degussa, 51, 73, 124, 126
Delta, 206
Department of Defense, 3, 79, 98, 118
Department of Energy, 1, 3, 7, 76, 79, 84
Department of Transportation, 76, 106, 219
deposition, 11, 16, 218
deposits, 223
depreciation, 19
depth, 1, 25, 36, 76, 155, 181, 183
designers, 228
destruction, 20
developed countries, 34, 79, 82, 122, 159, 217
developing countries, 87, 92
diesel fuel, 87, 101
digital cameras, 22, 24, 37, 82, 84, 118, 123, 128, 138, 140, 144, 146, 147, 225, 227, 229
digital communication, 156
disaster, 101
discs, 162
displacement, 91
distribution, 14, 49, 80, 87, 218, 220, 229
doctors, 138
domestic demand, 5, 23
domestic markets, 222

domestic resources, 144
DOT, 76, 106
draft, 54
drying, 164
durability, 116

E

Eastern Europe, 82, 109
economic development, 51, 93
economic growth, 95
economic reform, 93
economics, 31, 156
education, 140, 161
electric bikes, vii, 3, 4, 8, 10, 21, 22, 24, 79, 80, 81, 82, 84, 86, 88, 89, 94, 95, 96, 97, 98, 99, 100, 102, 103, 106, 108, 109, 110, 112, 116, 118, 120, 125, 133, 139, 140, 142, 143, 146, 157, 163, 201, 207, 216, 217, 227
electric vehicles (EVs), vii, 3, 10, 80, 200
electricity, 14, 52, 80, 88, 91, 94, 95, 102, 104
electrochemistry, 16
electrodes, 10, 12, 16, 20, 52, 56, 126, 168
electrolyte, 2, 6, 10, 11, 12, 16, 19, 20, 21, 24, 25, 33, 44, 45, 127, 128, 129, 130, 131, 155, 223
emergency, 30, 51, 223
emission, 67, 81, 82, 87, 102, 108, 117, 181, 200
employees, 4, 22, 30, 38, 46, 51, 71, 117, 121, 138, 139, 141, 144, 151, 163, 200, 219, 220, 228
encapsulation, 225, 226
encouragement, 2, 79
endothermic, 15
energy, 3, 4, 6, 7, 8, 9, 10, 11, 12, 14, 15, 16, 17, 18, 19, 20, 25, 29, 30, 31, 33, 35, 36, 39, 45, 47, 48, 51, 52, 53, 54, 55, 56, 62, 67, 68, 71, 80, 82, 87, 89, 90, 91, 94, 96, 100, 102, 103, 105, 111, 114, 115, 117, 118, 122, 123, 124, 125, 128, 135, 136, 137, 139, 145, 151, 153, 155, 157, 159, 165, 169, 179, 181, 182, 183, 199, 202, 214, 216, 217, 218, 219, 220, 222, 223, 224, 226, 227
energy consumption, 30, 56, 67, 87, 89, 96, 102, 181, 223, 224
energy density, 6, 11, 29, 30, 31, 48, 51, 123, 124, 125, 128, 145, 155, 157, 216, 223
energy efficiency, 7, 14, 16, 55, 179
energy supply, 53, 117
engineering, 20, 24, 43, 44, 56, 75, 115, 144, 145, 152, 157, 207, 224, 228
environment, 3, 12, 20, 22, 42, 48, 108, 113, 117, 125, 157, 160, 162, 165, 168, 217
environmental impact, 102
environmental issues, 116
environmental protection, 53, 105, 163, 219, 220

Environmental Protection Agency, 76, 116, 172, 203
environmental quality, 116
environmental regulations, 91
EPA, 76, 116, 117, 172, 203
EPS, 210
equality, 229
equipment, 4, 6, 19, 21, 24, 30, 34, 35, 37, 39, 40, 43, 46, 47, 53, 55, 71, 77, 82, 83, 118, 120, 121, 122, 131, 134, 135, 141, 142, 147, 150, 159, 162, 163, 164, 167, 199, 202, 205, 206, 217, 218, 221, 228, 229
equity, 71, 200
ethanol, 216
ethylene, 76, 130, 131
EU, 76, 81, 87, 97, 102, 122, 142, 162, 165
Europe, 5, 9, 10, 21, 26, 37, 80, 81, 82, 86, 92, 98, 105, 109, 112, 118, 126, 139, 144, 162, 164, 171, 219, 220, 221, 222, 229
European Commission, 139
European market, 107
European Parliament, 109
European Union, 76, 81, 87, 109, 217
evidence, 153
evolution, 21
excitation, 46, 134
exercise, 110, 162, 205
exertion, 95
expertise, 224
export market, 4, 21, 107, 121
exporter, 113, 114, 115, 117, 204, 205, 206, 207, 208, 210, 211, 212, 215
exporters, 83
exports, 81, 83, 92, 97, 104, 110, 126, 160
extraction, 133

F

fabrication, 12, 16, 145
factories, 93, 228
families, 93
farmland, 101
fears, 100
federal government, 40, 44, 46, 52
Federal Government, 81, 171
fiber, 110, 124, 164
field trials, 112
films, 52
financial, 40, 108, 111, 214
financial crisis, 108
financial incentives, 111
Finland, 220
fires, 21
flame, 131

flexibility, 30, 223
flexible manufacturing, 227
flight, 226
fluid, 12, 13
FMC, 133
force, 16, 78, 87, 116, 126
Ford, 59, 82, 106
foreign companies, 7, 22, 56, 84, 121
foreign investment, 100
formation, 16, 34, 93
foundations, 34
France, 32, 59, 122, 138, 159, 172, 217
friction, 94
frost, 60
fuel cell, vii, 3, 5, 6, 28, 29, 35, 36, 43, 44, 48, 52, 76, 116, 118, 136, 137, 202, 208, 212, 216
fuel cell vehicles, vii, 3, 29, 44, 118, 212
fuel consumption, 88
fuel efficiency, 32, 88
fuel prices, 108
funding, 21, 40, 44, 55, 146
funds, 166

G

garbage, 162
gasification, 52
GDP, 76, 80, 87, 172
gel, 112, 155
General Motors, 1, 10, 29, 43, 76, 118
geometry, 14
Georgia, 214
Germany, 42, 47, 57, 59, 80, 81, 82, 87, 92, 97, 109, 110, 122, 124, 135, 146, 159, 211, 217, 229
global competition, 145
global economy, 151
global warming, 112
glycol, 13
government policy, 82, 115
governments, 44, 47, 135
grades, 10
grading, 35
grants, 5, 29, 32, 34, 46, 96
graphite, 6, 10, 19, 20, 24, 45, 119, 128
gravity, 78, 116
Great Britain, 81, 97
greenhouse, 76, 87
gross domestic product, 76, 80, 87
growth, 4, 9, 16, 22, 23, 24, 41, 45, 48, 52, 80, 81, 82, 86, 87, 90, 92, 93, 97, 99, 101, 104, 105, 118, 121, 122, 130, 132, 136, 139, 145, 158, 160, 164, 217, 222, 229

growth rate, 22, 24, 48, 80, 81, 93, 101, 104, 130, 139
Guangdong, 50, 104, 140, 177
Guangzhou, 61, 72, 73, 80, 89, 90, 96, 99, 101, 105, 131, 132, 161, 165, 171, 172
guidance, 225

H

HE, 17
health, 31
heat removal, 14
heat transfer, 12, 13
height, 32, 68, 116
highways, 101
hiring, 52
history, 26, 97, 99, 108, 132
hobby, 205
holding company, 49, 75, 144
homes, 211
Hong Kong, 37, 51, 72, 73, 107, 124, 138, 141, 142, 144, 176, 211, 221, 222
host, 10, 16
House, 76, 106, 205
House of Representatives, 76, 106
housing, 138
hub, 81, 88, 98, 99, 110, 117, 163, 207
human, 80, 88, 91, 92, 95, 103, 105
Hungary, 81, 97
hybrid, vii, 1, 2, 3, 5, 8, 9, 10, 12, 21, 24, 28, 29, 32, 33, 36, 37, 39, 41, 43, 46, 49, 50, 52, 53, 56, 71, 76, 77, 82, 83, 84, 85, 117, 118, 119, 125, 132, 137, 200, 201, 202, 203, 211, 212, 213, 215, 216, 227
hybrid vehicles, vii, 3, 5, 8, 10, 12, 21, 24, 28, 29, 41, 43, 46, 52, 56, 71, 82, 118, 119, 203
hydrogen, 5, 44, 52, 94, 212, 214
hydrogen gas, 52

I

ideal, 31, 157, 216, 219
image, 104, 226, 229
imbalances, 12
imported products, 129
imports, 6, 45, 87, 101, 104, 122, 129, 165
improvements, 22, 56, 88, 115, 168, 220, 223
impurities, 130
incidence, 120
income, 55, 80, 87, 93, 151, 167
income tax, 55, 167

India, 82, 92, 105, 107, 111, 112, 113, 114, 115, 169, 206
individuals, 9
Indonesia, 92, 107
industrial sectors, 127
industrialization, 30, 52, 53, 55, 87, 157, 220, 222, 223
industrialized countries, 88
industries, 4, 29, 30, 34, 85, 93, 116, 121, 136, 168, 202, 211, 220, 222, 223
industry, 4, 5, 6, 7, 8, 9, 10, 18, 21, 22, 23, 24, 29, 30, 43, 44, 45, 48, 50, 52, 53, 56, 57, 81, 83, 85, 86, 87, 93, 95, 97, 98, 99, 100, 101, 105, 109, 110, 114, 116, 117, 119, 121, 122, 123, 125, 126, 127, 130, 134, 136, 138, 139, 142, 145, 146, 151, 153, 158, 159, 161, 163, 165, 166, 168, 217, 221, 222, 223, 224, 226, 229
infancy, 93
information exchange, 143
information technology, 7, 47, 76, 122, 222
infrastructure, 80, 87, 88, 92, 116, 141, 224
initiation, 7, 53
injuries, 120
institutions, 27, 57, 146
integration, 152, 157
integrity, 16, 20
intellectual property, 37, 71, 81, 99, 122, 145, 200, 220, 221, 222, 224
intellectual property rights, 37, 145, 220, 221, 222
interest groups, 100
interface, 2, 10, 129
interference, 100
internal controls, 142
international standards, 128, 163
international trade, 161
internship, 42, 146
interphase, 131
investment, 6, 7, 36, 37, 49, 50, 51, 52, 97, 126, 144, 220, 221, 226, 228
investment incentive, 52
investors, 55, 100, 167
ions, 10, 11, 12, 129
Iowa, 215
IPR, 76
Ireland, 209
iron, 2, 11, 15, 18, 54, 93, 94, 138, 215
issues, 8, 21, 56, 66, 96, 106, 109, 124, 137, 139, 158, 159, 165, 168, 218, 224
Italy, 81, 82, 97, 109

J

Japan, 3, 5, 6, 7, 10, 21, 23, 24, 26, 28, 32, 35, 37, 40, 45, 47, 48, 58, 79, 80, 81, 82, 83, 87, 92, 97, 101, 105, 107, 108, 109, 112, 118, 119, 121, 122, 123, 125, 126, 129, 135, 137, 159, 162, 163, 171, 172, 217, 227, 229
Java, 211
joint ventures, 4, 5, 7, 8, 21, 26, 49, 56, 84, 85, 121, 200
justification, 100

K

Korea, 3, 23, 32, 37, 79, 118, 121, 122, 222, 229

L

lakes, 5, 24, 133, 218
Laos, 107
laptop, 18, 21, 23, 140, 220, 223, 227
laws, 55, 109, 110, 145, 167
laws and regulations, 55, 167
lead, vii, 2, 3, 12, 14, 21, 26, 29, 30, 35, 50, 51, 54, 77, 79, 81, 82, 83, 84, 86, 90, 91, 92, 94, 95, 97, 99, 100, 102, 107, 112, 117, 118, 120, 124, 132, 133, 134, 137, 138, 140, 142, 151, 157, 159, 160, 165, 167, 206, 212, 213, 215, 216, 220, 223
leadership, 144
leakage, 42
LED, 76, 114, 225, 226
legislation, 100, 102, 104
leisure, 110, 164
lens, 113
liability insurance, 164
lifetime, 89, 92
light, 53, 56, 76, 81, 82, 91, 100, 102, 108, 114, 116, 127, 162, 200
Lion, 52, 73, 126
liquids, 13
lithium, vii, 1, 2, 3, 4, 5, 6, 7, 8, 9, 10, 11, 12, 14, 15, 16, 17, 18, 19, 20, 21, 22, 23, 24, 25, 26, 27, 28, 29, 30, 31, 32, 33, 34, 35, 36, 37, 38, 39, 41, 43, 44, 45, 46, 47, 48, 49, 51, 52, 53, 54, 56, 57, 61, 62, 66, 71, 76, 77, 79, 84, 124, 128, 130, 133, 155, 179, 199, 212, 215, 216, 217, 218, 219, 220, 221, 222, 223, 224, 225, 226, 227, 228, 229
lithium ion batteries, 212
Lithium-ion battery technology, vii, 3
local authorities, 111
local government, 30, 35, 89, 93, 165, 223
logistics, 2

low temperatures, 124, 168, 179
LPG, 76, 89, 91, 103
LTD, 73

M

machinery, 94, 113, 141
magnesium, 76, 133
magnet, 46, 47, 134, 135
magnetic field, 47, 135
magnitude, 15
Mainland China, 50, 51, 107
majority, 80, 88, 92, 107
Malaysia, 92, 107, 229
management, 3, 4, 5, 9, 12, 13, 14, 15, 19, 29, 31, 32, 33, 37, 44, 66, 69, 80, 86, 90, 97, 111, 115, 124, 137, 141, 142, 144, 145, 146, 156, 158, 162, 164, 165, 179, 180, 182, 217, 220, 222, 224, 225, 226, 227
manganese, 2, 11, 30, 77, 126, 223
manpower, 113
manufacturing, 4, 5, 6, 7, 8, 9, 10, 20, 21, 24, 26, 30, 37, 39, 46, 47, 48, 49, 51, 52, 55, 56, 71, 83, 84, 85, 86, 97, 99, 109, 111, 114, 117, 118, 120, 124, 129, 131, 133, 134, 138, 140, 141, 142, 144, 147, 157, 162, 163, 167, 168, 199, 220, 224, 225, 226, 227, 229
manufacturing companies, 5, 133, 168
market economy, 93
market penetration, 111
market share, 22, 23, 37, 48, 91, 92, 110, 122
marketability, 8, 85
marketing, 6, 9, 30, 33, 37, 38, 51, 86, 95, 106, 108, 141, 143, 201, 222, 227, 228, 229
marketplace, 94, 116, 227
mass, 13, 24, 35, 47, 48, 49, 67, 80, 86, 87, 94, 117, 124, 125, 126, 129, 133, 135, 137, 139, 143, 144, 155, 160, 181
materials, 5, 6, 7, 9, 10, 11, 16, 19, 20, 24, 29, 30, 32, 33, 35, 36, 45, 46, 48, 52, 53, 66, 83, 84, 85, 105, 118, 123, 124, 126, 128, 129, 130, 131, 136, 141, 142, 145, 159, 163, 179, 217, 218, 222, 223, 224, 225
materials science, 223
matter, 77, 100, 102
measurement, 34, 201
measurements, 54, 142
media, 13, 110, 111
medical, 6, 39, 49, 52, 147, 159, 218, 225, 227
melting, 130
membership, 5, 29
membranes, 12
memory, 30, 31, 52, 179, 223

mergers, 165
metallurgy, 45
metals, 6, 45, 145
meter, 78, 130, 156, 163
methanol, 52
methodology, 71, 199
metropolitan areas, 116
Mexico, 211
Middle East, 139, 162, 164
military, 35, 98, 159, 217, 219, 225
mineral resources, 29
Missouri, 207
mixing, 40
mobile communication, 30, 223, 229
mobile phone, 21, 22, 23, 24, 30, 37, 48, 49, 54, 84, 121, 127, 138, 144, 223, 225, 227
models, 80, 86, 91, 94, 96, 109, 110, 126, 133, 137, 143, 147, 160, 165, 220, 225
modifications, 36
modules, 5, 13, 14, 15, 32, 33, 34, 35, 54, 138, 157, 211
moisture, 130, 131
moisture content, 130
molecules, 130
Mongolia, 26
monopoly, 6, 29, 136, 222
motivation, 83
motor control, 100, 135
music, 205

N

nanomaterials, 50
nanometer, 78
nanotechnology, 6, 45, 49
national policy, 132, 165
national product, 81, 97
natural resources, 113
NCA, 11, 20
negotiating, 51
Nepal, 107
Netherlands, 80, 82, 92, 109, 170
neutral, 96
New Zealand, 170, 229
next generation, 113
nickel, 2, 3, 5, 11, 14, 22, 24, 26, 30, 34, 35, 48, 50, 54, 77, 80, 84, 94, 105, 121, 216, 217, 220, 223, 228, 229
nitrogen, 52, 77
North America, 14, 37, 67, 144, 180, 219, 220, 221, 222, 229

O

obstacles, 23
Oceania, 229
officials, 23, 83, 99, 120
oil, 8, 9, 13, 78, 86, 87, 101, 103, 125, 165
operating range, 4, 15, 71, 199
operations, 44, 48, 55, 99, 136, 167, 222
opportunities, 34
optimization, 53, 137
Organization for Economic Cooperation and Development, 77, 87
ownership, 80, 81, 87, 92, 93, 94, 101
oxygen, 52

P

Pacific, 106, 176, 208, 213, 215
Pakistan, 107
parallel, 14, 15, 18, 31, 46, 66, 67, 77, 84, 134, 140, 157, 158, 180, 217
Parliament, 109
patents, 5, 29, 32, 33, 37, 38, 40, 46, 47, 71, 83, 84, 123, 129, 135, 137, 140, 147, 153, 164, 200, 219, 224
PCM, 77, 143, 144, 147
pedal, 76, 94, 96, 98, 99, 100, 109, 201
performers, 223
personal benefit, 80, 87
personal computers, 49
petroleum, 76, 87, 88, 89, 101
Petroleum, 103
Philippines, 107
phosphate, 2, 11, 15, 18, 49, 124, 215
photovoltaic cells, 114
photovoltaic devices, 54
platform, 116, 153, 156, 157
playing, 80, 87, 122, 129, 130, 164
plug-in hybrid vehicles (PHEVs), vii, 3
PM, 77, 102, 103, 107
polarization, 129, 147
police, 90
policy, 7, 55, 96, 116, 165, 220
policymakers, 96
pollution, 7, 30, 52, 55, 82, 87, 96, 99, 100, 105, 216, 223, 224
polymer, 2, 23, 26, 36, 44, 48, 49, 51, 61, 77, 84, 125, 139, 140, 141, 143, 144, 146, 147, 152, 153, 155, 157, 212, 216, 219, 221, 224, 227, 229
polymers, 227
polypropylene, 6, 12, 124
poor performance, 14

population, 21, 87, 100, 111
population density, 87
portable computers, vii, 3, 7, 22, 47, 118, 223
portable electronics products, vii, 3
portfolio, 145
PRC, vii, 2, 4, 7, 8, 9, 21, 28, 51, 52, 55, 71, 76, 77, 79, 81, 83, 84, 86, 101, 102, 103, 120, 136, 138, 140, 167, 171, 173
president, 45
President, 44, 64
primary cells, 50
primary school, 161
principles, 54, 221, 229
privatization, 98
process control, 56, 168
producers, 56, 133, 168, 229
product market, 22
professionals, 138, 143, 228
profit, 151
project, 24, 38, 44, 48, 50, 79, 97, 114, 117, 123, 153, 161, 166, 207, 212, 214, 220, 221, 228
promotion campaigns, 110
propane, 130
prosperity, 101
protection, 21, 22, 23, 25, 31, 32, 66, 81, 99, 113, 127, 179, 220, 224
prototype, 13, 17, 36, 82, 99, 100, 111, 157
public domain, 5
public safety, 81, 98
purity, 19, 130, 131

Q

qualifications, 142
quality assurance, 227
quality control, 6, 29, 71, 77, 124, 200, 224, 227
quality of life, 160

R

radiation, 32
radio, 138, 143, 147, 205
raw materials, 19, 24, 97, 122, 141, 142, 165, 199, 218
reactions, 20
reactivity, 10
real estate, 29, 136, 222
recognition, 108, 109
recommendations, 8, 56, 168
recreation, 92
recreational, 92, 205, 213
recycling, 85, 90, 91

reform, 93, 145
Reform, 48, 52, 53
regulations, vii, 48, 55, 79, 85, 87, 96, 106, 108, 112, 117, 138, 142
rejection, 13
reliability, 6, 8, 9, 16, 18, 21, 22, 29, 30, 56, 84, 85, 116, 124, 125, 136, 142, 153, 168, 222, 223, 227
renewable energy, 7, 52, 53, 55, 122
repair, 95, 114, 202, 204, 206, 207, 208, 212, 214
reputation, 126, 142, 163, 219
requirements, 3, 6, 10, 16, 17, 35, 36, 41, 42, 45, 85, 90, 106, 109, 115, 118, 120, 127, 130, 131, 142, 146, 151, 224
research institutions, 229
researchers, 12, 42, 125, 146, 228
resistance, 12, 15, 16, 94, 129, 131
resources, 5, 24, 25, 30, 34, 49, 52, 106, 122, 223, 224
response, 3, 12, 20, 54, 80
retail, 49, 71, 101, 110, 114, 138, 200, 204, 205, 206, 207, 208, 210, 212, 215, 216
revenue, 48, 132, 134, 139, 160
rights, 71, 165, 200
risks, 20
room temperature, 14, 15
routes, 157
routines, 151
Royal Society, 170
royalty, 51
rules, 54

S

safety, 14, 15, 16, 18, 19, 20, 21, 30, 31, 32, 35, 41, 49, 54, 56, 66, 80, 83, 89, 90, 91, 94, 95, 96, 99, 106, 113, 120, 123, 124, 125, 126, 127, 128, 129, 131, 147, 151, 155, 156, 158, 159, 165, 168, 180, 216, 217, 223, 224, 225, 228
salts, 19
Samsung, 20, 22, 37, 121, 122, 125, 222
saturation, 93
savings, 96, 105
scaling, 56, 168
school, 6, 44, 93
science, 34, 145, 153, 222
scientific knowledge, 163
scope, 8, 223, 224
security, 51, 52, 53, 87, 205, 224, 225
selenium, 77, 128
semiconductor, 45, 54
seminars, 9, 86
sensitivity, 3, 12, 14, 16
sensors, 21, 35, 52

service provider, 203
services, 44, 100, 117, 138, 142, 145, 147, 204, 229
shape, 155, 157, 216, 224
shock, 94
shortage, 87, 108
showing, 121
signals, 116
silicon, 13, 35, 54, 77, 128
silver, 35, 54
simulation, 36, 157
Singapore, 60, 107
social development, 34
social responsibility, 145
society, 5, 9, 29, 93, 105, 161, 163
software, 34, 43
solar cells, 35, 54, 114, 227
solar collectors, 114
solid oxide fuel cells, 52
solid waste, 90
solution, 12, 109, 155, 224
solvents, 10, 16, 130, 131
South Africa, 204
South America, 162
South Korea, 58, 59
Southeast Asia, 91, 92, 139, 144, 162, 221
Spain, 210
specialists, 146
species, 20
specifications, 12, 37, 54, 56, 136, 140, 143, 144, 151, 157, 165, 180, 221
speech, 98, 108
Sri Lanka, 107
stability, 11, 116, 128, 159, 218
staff members, 34, 44, 136, 141
stakeholders, 10
standardization, 69, 182
state, 2, 8, 11, 12, 15, 16, 18, 21, 30, 31, 35, 40, 44, 47, 51, 105, 125, 135, 138, 144, 145, 151, 153, 164, 219, 220, 221, 223, 224
states, 15, 29, 106
statistics, 50, 83, 97, 105, 110
steel, 31, 32, 41, 93, 98, 115, 220
stock exchange, 144
storage, 5, 6, 10, 11, 29, 30, 34, 44, 50, 51, 52, 80, 88, 128, 136, 159, 166, 179, 217, 218, 222, 223
stress, 16, 157
stroke, 91, 99, 117
structure, 46, 52, 53, 115, 128, 129, 134, 157, 220
style, 75, 77, 80, 88, 91, 92, 93, 96, 97, 113, 145, 156
submarines, 219
subsidy, 111, 115
sulfur, 52, 87
Sun, 58, 63, 73, 146, 175

superconductor, 45
supervision, 34, 35
supplier, 49, 50, 113, 114, 118, 202, 204, 205, 206, 207, 208, 209, 210, 211, 213, 214, 215, 216, 224, 229
suppliers, 3, 9, 12, 79, 86, 110, 142, 146, 169
supply chain, 10, 122
surging, 108
sustainability, 170
sustainable development, 53, 105, 122, 159
Sweden, 206
swelling, 131
Switzerland, 82, 109, 110, 111
synthesis, 6, 29, 30, 31, 36, 84, 136, 222, 223

T

Taiwan, 37, 77, 82, 90, 92, 94, 105, 107, 109, 115, 116, 117, 118, 124, 132, 144, 157, 170, 171, 172, 212, 216, 221
talent, 138
tanks, 91
target, 126, 129, 142, 157, 161, 163
tax incentive, 55, 167
taxes, 55, 167
taxis, 101, 136
teams, 26, 144
technical assistance, 40
technical support, 138
techniques, 5, 10, 37, 48, 118, 141, 143, 152, 164
technological advances, 141
technological progress, 105, 162
telecommunications, 83
telephone, 140, 220
telephones, 27, 146, 147, 227
temperature, 3, 4, 11, 12, 13, 14, 15, 16, 19, 20, 21, 31, 33, 42, 53, 54, 56, 66, 67, 69, 71, 83, 124, 127, 128, 130, 131, 143, 145, 147, 159, 166, 181, 182, 199, 217, 225
terminals, 54
test data, 5, 155
testing, 4, 6, 8, 9, 20, 21, 22, 28, 30, 31, 32, 34, 35, 38, 39, 40, 42, 43, 45, 46, 51, 53, 56, 71, 83, 85, 86, 116, 126, 127, 134, 136, 137, 138, 139, 141, 146, 150, 157, 163, 199, 212, 224, 227
Thailand, 92, 107, 208
thermal decomposition, 11
thermal stability, 30, 223
three-wheeled vehicles, vii, 79, 82, 85, 98, 111, 160
Tibet, 133, 218
titanate, 50
titanium, 145
tooth, 225

total energy, 7, 55, 115, 123
total revenue, 134
tourism, 29, 136, 222
Toyota, 13, 14, 20, 43, 68, 126, 182, 203
toys, 84, 126, 138, 144, 147, 159, 205, 218, 225, 227
tracks, 46, 106, 134
trade, 110, 162, 221
trading partners, 45
training, 110, 150, 200
transition period, 93
transmission, 115
transport, 12, 89, 93, 96, 97, 102, 110, 120, 164
transportation, vii, 3, 27, 51, 80, 81, 82, 87, 88, 89, 92, 94, 95, 98, 100, 101, 102, 104, 105, 113, 116, 118, 120, 126, 135, 143, 159, 160, 164, 165, 169, 212, 216, 217, 219, 224
transportation applications, vii, 3, 27, 118
treatment, 36, 55, 124, 167
trial, 6, 30, 35, 136, 152, 222, 223
troubleshooting, 151
twist, 201

U

uniform, 12
unit cost, 56, 141, 168
United Kingdom, 170, 204, 205
United Nations, 35
United States (USA), vii, 3, 5, 6, 8, 9, 26, 32, 34, 35, 37, 39, 42, 44, 45, 46, 47, 56, 57, 58, 71, 79, 81, 82, 85, 86, 92, 97, 98, 101, 106, 107, 112, 118, 122, 125, 129, 135, 139, 146, 147, 159, 162, 164, 168, 200, 217, 219, 220
universities, 8, 27, 37, 42, 57, 86, 146
urban, 10, 82, 88, 89, 90, 92, 93, 94, 95, 96, 97, 101, 102, 105, 118, 201, 224
urban areas, 88, 89, 93, 102
urban population, 93

V

valve, 2, 14, 77, 82, 92, 147
vanadium, 124
vapor, 130

variations, 12
velocity, 13, 47, 135
vibration, 83, 116, 127
Vice President, 2, 65, 175, 178
victims, 83, 120
videos, 23
Vietnam, 92, 107, 172
viscosity, 13, 130
Volkswagen, 43

W

walking, 93, 102
Washington, 58, 59, 60, 206, 211
waste, 90
watches, 205
water, 10, 13, 53, 114
water heater, 114
web, 26, 110, 138, 140, 142, 216
web sites, 26
welding, 115, 164
Western countries, 100, 112
wholesale, 113, 114, 118, 138, 202, 204, 205, 206, 207, 208, 209, 210, 211, 213, 214, 215, 216
wind turbines, 115, 210, 211
Wisconsin, 125, 126
workers, 7, 48, 161
workforce, 81, 99
World Bank, 226
World Trade Organization, 163
worldwide, vii, 3, 7, 21, 37, 49, 52, 53, 56, 92, 95, 96, 107, 118, 125, 168, 219, 220, 222, 229

Y

yield, 19
young people, 111
yuan, 50, 51, 78, 81, 86, 200

Z

zinc, 35, 54, 105